W9-BRA-180

STATISTICAL THERMODYNAMICS

STATISTICAL THERMODYNAMICS

Erwin Schrödinger

Dover Publications, Inc.
New York

This Dover edition, first published in 1989, is an unabridged and unaltered republication of the second edition (1952) of the work first published in 1946 by the Cambridge University Press, Cambridge, England with the following subtitle: *A Course of Seminar Lectures Delivered in January-March 1944, At the School of Theoretical Physics, Dublin Institute For Advanced Studies.*

Library of Congress Cataloging-in-Publication Data

Schrödinger, Erwin, 1887-1961.
Statistical thermodynamics / Erwin Schrödinger.
p. cm.
Reprint. Originally published: 2nd ed. Cambridge : Cambridge University Press, 1952.
ISBN-13: 978-0-486-66101-8
ISBN-10: 0-486-66101-6
1. Statistical thermodynamics. I. Title.
QC311.5.S36 1989
536′.71—dc20

Manufactured in the United States by Courier Corporation
66101611 2014
www.doverpublications.com

CONTENTS

NOTE

A very small edition of these Lectures was published in hectograph form by the Dublin Institute for Advanced Studies. It is hoped that the present edition, for which the text has been slightly revised, may reach a wider circle of readers.

NOTE ON SECOND EDITION

The view that a physical process consists of continual jump-like transfers of energy parcels between microsystems cannot, when given serious thought, pass for anything but a sometimes convenient metaphor. To ascribe to every system always one of its sharp energy values is an indefensible attitude. It was challenged in the beginning of Chapter II, yet it was adopted throughout this treatise as a customary convenient short-cut. The Appendix added to the Second Edition contains the general proof, that a consistent procedure, based on very simple assumptions, always gives the same results. The thermodynamical functions depend on the quantum-mechanical level-scheme, not on the gratuitous allegation that these levels are the only allowed states.

Except for the Appendix the Second Edition is a reprint.

E. S.

June, 1952

CHAPTER I

GENERAL INTRODUCTION

THE object of this seminar is to develop briefly one simple, unified standard method, capable of dealing, without changing the fundamental attitude, with all cases (classical, quantum, Bose-Einstein, Fermi-Dirac, etc.) and with every new problem that may arise. The interest is focused on the general procedure, and examples are dealt with as illustrations thereof. It is not a first introduction for newcomers to the subject, but rather a 'repetitorium'. The treatment of those topics which are to be found in every one of a hundred text-books is severely condensed; on the other hand, vital points which are usually passed over in all but the large monographs (such as Fowler's and Tolman's) are dealt with at greater length.

There is, essentially, only one problem in statistical thermodynamics: the distribution of a given amount of energy E over N identical systems. Or perhaps better: to determine the distribution of an assembly of N identical systems over the possible states in which this assembly can find itself, given that the energy of the assembly is a constant E. The idea is that there is weak interaction between them, so weak that the energy of interaction can be disregarded, that one can speak of the 'private' energy of every one of them and that the sum of their 'private' energies has to equal E. The distinguished role of the energy is, therefore, simply that it is a constant of the motion—the one that always exists, and, in general, the only one. The generalization to the case, that there are others besides (momenta, moments of momentum), is obvious; it has occasionally been contemplated, but in terrestrial, as opposed to astrophysical, thermodynamics it has hitherto not acquired any importance.

'To determine the distribution' means in principle to make oneself familiar with any possible distribution-of-the-energy (or state-of-the-assembly), to classify them in a suitable way, i.e. in the way suiting the purpose in question and to count the numbers in the classes, so as to be able to judge of the probability of certain features or characteristics turning up in the assembly. The questions that can arise in this respect are of the most varied nature, and so the classification really needed in a special problem can be of the most varied nature, especially in relation to the fineness of classification. At one end of the scale we have the general question of finding out those features which are common to almost all possible states of the assembly so that we may safely contend that they 'almost always' obtain. In this case we have well-nigh only one class—actually two, but the second one has a negligibly small content. At the other end of the scale we have such a detailed question as: volume (= number of states of the assembly) of the 'class' in which one individual member is in a particular one of its states. Maxwell's law of velocity distribution is the best-known example.

This is the mathematical problem—always the same; we shall soon present its general solution, from which in the case of every particular kind of system every particular classification that may be desirable can be found as a special case.

But there are two different attitudes as regards the physical application of the mathematical result. We shall later, for obvious reasons, decidedly favour one of them; for the moment, we must explain them both.

The older and more naïve application is to N actually existing physical systems in actual physical interaction with each other, e.g. gas molecules or electrons or Planck oscillators or degrees of freedom ('ether oscillators') of a 'hohlraum'. The N of them together represent the actual physical system under consideration. This original point of view is associated with the names of Maxwell, Boltzmann and others.

But it suffices only for dealing with a very restricted class of physical systems—virtually only with gases. It is not applicable to a system which does not consist of a great number of identical constituents with 'private' energies. In a solid the interaction between neighbouring atoms is so strong that you cannot mentally divide up its total energy into the private energies of its atoms. And even a 'hohlraum' (an 'ether block' considered as the seat of electromagnetic-field events) can only be resolved into oscillators of many—infinitely many—different types, so that it would be necessary at least to deal with an assembly of an infinite number of different assemblies, composed of different constituents.

Hence a second point of view (or, rather, a different application of the same mathematical results), which we owe to Willard Gibbs, has been developed. It has a particular beauty of its own, is applicable quite generally to every physical system, and has some advantages to be mentioned forthwith. Here the N identical systems are mental copies of the one system under consideration—of the one macroscopic device that is actually erected on our laboratory table. Now what on earth could it mean, physically, to distribute a given amount of energy E over these N mental copies? The idea is, in my view, that you can, of course, imagine that you really had N copies of your system, that they really were in 'weak interaction' with each other, but isolated from the rest of the world. Fixing your attention on one of them, you find it in a peculiar kind of 'heat-bath' which consists of the $N-1$ others.

Now you have, on the one hand, the experience that in thermodynamical equilibrium the behaviour of a physical system which you place in a heat-bath is always the same whatever be the nature of the heat-bath that keeps it at constant temperature, provided, of course, that the bath is chemically neutral towards your system, i.e. that there is nothing else but heat exchange between them. On the other hand, the statistical calculations do not refer to the mechanism of interaction; they only assume that it is 'purely mechanical', that it does not affect the nature of the single systems (e.g. that it never blows them to pieces), but merely transfers energy from one to the other.

These considerations suggest that we may regard the behaviour of any one of those N systems as describing the *one* actually existing system when placed in a heat-bath of given temperature. Moreover, since the N systems are alike and under similar conditions, we can then obviously, from their simultaneous statistics, judge of the probability of finding our system, when placed in a heat-bath of given temperature, in one or other of its private states. Hence all questions concerning the system in a heat-bath can be answered.

We adopt this point of view in principle—though all the following considerations may, with due care, also be applied to the other. The advantage consists not only in the general applicability, but also in the following two points:

(i) N can be made arbitrarily large. In fact, in case of doubt, we always mean $\lim N = \infty$ (infinitely large heat-bath). Hence the applicability, for example, of Stirling's formula for $N!$, or for the factorials of 'occupation numbers' proportional to N (and thus going with N to infinity), need never be questioned.

(ii) No question about the individuality of the members of the assembly can ever arise—as it does, according to the 'new statistics', with particles. Our systems are macroscopic systems, which we could, in principle, furnish with labels. Thus two states of the assembly differing by system No. 6 and system No. 13 having exchanged their roles are, of course, to be counted as different states—while the same may not be true when two similar atoms within system No. 6 have exchanged their roles; but the latter is merely a question of enumerating correctly the states of the single system, of describing correctly its quantum-mechanical nature.

CHAPTER II

THE METHOD OF THE MOST PROBABLE DISTRIBUTION

WE are faced with an assembly of N identical systems. We describe the nature of any one of them by enumerating its possible states, which we label $1, 2, 3, 4, \ldots, l, \ldots$. In principle we have always in mind a quantum-mechanical system whereby the states are to be described by the eigenvalues of a complete set of commuting variables. The eigenvalues of the energy in these states we call $\epsilon_1, \epsilon_2, \epsilon_3, \ldots, \epsilon_l, \ldots$, ordered so that $\epsilon_{l+1} \geqslant \epsilon_l$. But, if necessity arises, the scheme can also be applied to a 'classical system', when the states will have to be described as cells in phase-space (p_k, q_k) of equal volume and—whether infinitesimal in all directions or not—at any rate such that the energy does not vary appreciably within a cell. More important than this merely casual application is the following:

We shall always regard the state of the assembly as determined by the indication that system No. 1 is in state, say, l_1, No. 2 in state $l_2, \ldots$, No. N in state l_N. We shall adhere to this, though the attitude is altogether wrong. For, a quantum-mechanical system is not in this or that state to be described by a complete set of commuting variables chosen once and for all. To adopt this view is to think along severely 'classical' lines. With the set of states chosen, the individual system can, at best, be relied upon as having a certain probability amplitude, and so a certain probability, of being, on inspection, found in state No. 1 or No. 2 or No. 3, etc. I said: at best a probability amplitude. Not even that much of determination of the single system need there be. Indeed, there is no clear-cut argument for attributing to the single system a 'pure state' at all.

If we were to enter on this argument, it would lead us far astray to very subtle quantum-mechanical considerations.

Von Neumann, Wigner and others have done so, but the results do not differ appreciably from those obtained from the simpler and more naïve point of view, which we have outlined above and now adopt.

Thus, a certain class of states of the assembly will be indicated by saying that $a_1, a_2, a_3, \ldots, a_l, \ldots$ of the N systems are in state $1, 2, 3, \ldots, l, \ldots$ respectively, and all states of the assembly are embraced—without overlapping—by the classes described by all different admissible sets of numbers a_l:

State No.	1	2	3	...	l	...	
Energy	ϵ_1	ϵ_2	ϵ_3	...	ϵ_l	...	(2·1)
Occupation No.	a_1	a_2	a_3	...	a_l	...	

The number of single states, belonging to this class, is obviously

$$P = \frac{N!}{a_1!\,a_2!\,a_3!\ldots a_l!\ldots}. \tag{2·2}$$

The set of numbers a_l must, of course, comply with the conditions

$$\sum_l a_l = N, \quad \sum_l \epsilon_l a_l = E. \tag{2·3}$$

The statements (2·2) and (2·3) really finish our counting. But in this form the result is wholly unsurveyable.

The present method admits that, on account of the enormous largeness of the number N, the total number of distributions (i.e. the sum of all P's) is very nearly exhausted by the sum of those P's whose number sets a_l do not deviate appreciably from that set which gives P its maximum value (among those, of course, which comply with (2·3)). In other words, if we regard this set of occupation numbers as obtaining always, we disregard only a very small fraction of all possible distributions —and this has 'a vanishing likelihood of ever being realized'.

This assumption is rigorously correct in the limit $N \to \infty$ (thus: in the application to the 'mental' or 'virtual' assembly, where *in dubio* we always mean this limiting case, which corresponds physically to an 'infinite heat-bath'; you see again the great superiority of the Gibbs point of view). Here we adopt this assumption without the proof which will emerge later from the alternative method—the Darwin-Fowler 'method of mean values'.

For N large, but finite, the assumption is only approximately true. Indeed, in the application to the Boltzmann case, the distributions with occupation numbers deviating from the 'maximum set' must not be entirely disregarded. They give information on the thermodynamic fluctuations of the Boltzmann system, when kept at constant energy E, i.e. in perfect heat isolation.

But we shall not work that out here, partly on account of the very restricted applicability of the Boltzmann point of view itself, and also for the following reason: Since the condition of perfect heat isolation cannot be practically realized, the results obtained for the thermodynamic fluctuations under this, non-realizable, condition apply to reality only in part, that is, in so far as they can be shown to be, or can be trusted to be, the same as 'under heat-bath condition'. Now the fluctuations of a system in a heat-bath at constant temperature are much more easily obtained directly from the Gibbs point of view. Hence there is no point in following up the more complicated device to obtain information which really only applies to an ideal, non-realizable, case.

Returning to (2·2) and (2·3), we choose the logarithm of P as the function whose maximum we determine, taking care of the accessory conditions in the usual way by Lagrange multipliers, λ and μ; i.e. we seek the unconditional maximum of

$$\log P - \lambda \sum_l a_l - \mu \sum_l \epsilon_l a_l; \tag{2·4}$$

for the logarithms of the factorials we use Stirling's formula in the form

$$\log(n!) = n(\log n - 1). \tag{2·5}$$

And, of course, we treat the a_l as though they were continuous variables. We get for the variation of (2·4)

$$-\sum_l \log a_l \delta a_l - \lambda \sum_l \delta a_l - \mu \sum_l \epsilon_l \delta a_l = 0,$$

and have to equate to nought the coefficients of every δa_l, thus (for every l)

$$\log a_l + \lambda + \mu\epsilon_l = 0,$$

or

$$a_l = e^{-\lambda-\mu\epsilon_l}.$$

λ and μ are to be determined from the accessory conditions, thus

$$\sum_l e^{-\lambda-\mu\epsilon_l} = N, \quad \sum_l \epsilon_l e^{-\lambda-\mu\epsilon_l} = E.$$

On dividing, member by member, we eliminate λ, but we can also obtain $e^{-\lambda}$ directly from the first formula. Calling $E/N = U$ the average share of energy of one system, we express our whole result thus:

$$\left.\begin{aligned} \frac{E}{N} &= U = \frac{\Sigma\epsilon_l e^{-\mu\epsilon_l}}{\Sigma e^{-\mu\epsilon_l}} = -\frac{\partial}{\partial\mu}\log\Sigma e^{-\mu\epsilon_l}, \\ a_l &= N\frac{e^{-\mu\epsilon_l}}{\Sigma e^{-\mu\epsilon_l}} = -\frac{N}{\mu}\frac{\partial}{\partial\epsilon_l}\log\Sigma e^{-\mu\epsilon_l}. \end{aligned}\right\} \qquad (2\cdot6)$$

The set of equations in the second line indicates the distribution of our N systems over their energy levels. It may be said to contain, in a nutshell, the whole of thermodynamics, which hinges entirely on this basic distribution. The relation itself is very perspicuous—the exponential $e^{-\mu\epsilon_l}$ indicates the occupation number a_l as a fraction of the total number N of systems, the sum in the denominator being only a 'normalizing factor'. But, of course, μ would have to be determined from the first equation as a function of the average energy U and the 'nature of the system' (i.e. the ϵ_l's); and, naturally, it is impossible to solve this equation generally with respect to μ. In fact, it is obvious that the functional dependence between μ and U is certainly not universal, but depends entirely on the nature of the system.

But very fortunately we can give to our relations a very satisfactory general physical interpretation, without solving that equation with respect to μ, because the latter (originally introduced just as a Lagrange multiplier, as a mathematical help) turns out to be a much more fundamental quantity than U; so much so, that the physicist is gratified to be given, in every particular case, U as a function of it, rather than vice versa, which would be quite unnatural.

To explain this without too many qualifications, we now definitively adopt the Gibbs point of view, namely, that we are dealing with a virtual ensemble, of which the single member is the system really under consideration. And since all the single members are of equal right, we may now, when it comes to physical interpretation, think of the a_l, or rather of the a_l/N, as the frequencies with which a single system, immersed in a large heat-bath, will be encountered in the state ϵ_l, while U is its average energy under these circumstances.

We now apply our results (2·6) to three different (assemblies of) systems, viz.

$$\left.\begin{array}{llll} & A & B & A+B, \\ \text{levels:} & \alpha_k & \beta_m & \epsilon_l = \alpha_k + \beta_m. \end{array}\right\} \tag{2·7}$$

By this we mean that in the first and in the second case the single members shall be any two different systems of the general type considered hitherto, while in the third case the single member shall consist of one system A and one system B put into loose energy contact, so that the general energy level in the third case is the sum of any α_k and any β_m (the index l standing really for the pair of indices (k, m)). A moment's consideration shows that in the third case the sum splits into a product of two sums, thus:

$$\sum_l e^{-\mu\epsilon_l} = \sum_k \sum_m e^{-\mu(\alpha_k+\beta_m)} = \sum_k e^{-\mu\alpha_k} \sum_m e^{-\mu\beta_m}. \tag{2·8}$$

Hence from (2·6)—always in this third case—the general occupation number a_l (which we may equivalently label $a_{(k,m)}$) reads

$$a_l \equiv a_{(k,m)} = N \frac{e^{-\mu(\alpha_k+\beta_m)}}{\sum_k e^{-\mu\alpha_k} \sum_m e^{-\mu\beta_m}}. \tag{2·9}$$

We continue to speak of the third case and inquire, what is the number of systems '$A+B$' with A on the particular level α_k? It is clearly found on summing (2·9) over all m. In the result the

$\Sigma e^{-\mu\beta}$ cancels in numerator and denominator and we are left with

$$\sum_m a_{(k,m)} = N \frac{e^{-\mu\alpha_k}}{\sum_k e^{-\mu\alpha_k}}.$$

It is thus seen that the entire statistical distribution of the A systems in the third case (including *inter multa alia* the mean value of their energy) is exactly the same as it would be in an A assembly (first case) provided that we arrange (by a suitable choice of E/N in the A case) for the value of μ to be the same in the two cases.

Since the same consideration applies to system B, we have, according to our interpretation, that if you put the systems A and B into loose contact with one another and put them in a heat-bath, each of them behaves exactly as it would when put into a heat-bath by itself, provided only that the three heat-baths are chosen so as to make the μ values equal in the three cases. In other words, if that is done, the established energy contact is idle and there is, on the average, no mutual influence or energy exchange.

This can hardly be interpreted otherwise than that equal μ means equal temperature. And since you can choose a standard system A once and for all ('thermometer') and put it into contact with any other system B, μ must be a universal function of the temperature T.

This conclusion will be considerably strengthened when we proceed to determine explicitly what function of the temperature μ is.

For this purpose it is well to take stock of an obvious but very important by-product of our preceding considerations. We have seen that in the '$A+B$' case the

$$\sum_l e^{-\mu\epsilon_l} = \sum_k e^{-\mu\alpha_k} \sum_m e^{-\mu\beta_m}.$$

Hence the function of μ, which we shall see to be very important, viz.

$$\log \sum_l e^{-\mu\epsilon_l} \tag{2·10}$$

(whose usefulness is clear from the last members of (2·6)) is additive for two systems in loose energy contact. That is the obvious, but relevant, statement to which I referred above.

Now what is the functional relation between μ and T? To tell the 'true' absolute temperature T from the lot of its monotonic functions $f(T)$, there is, as is well known, only one criterion: $1/T$ is a universal integrating factor of the infinitesimal heat supply dQ in thermodynamic equilibrium—universal, that is to say, for any system. No other function of T has this property—it is the definition of T (Kelvin).

To avail ourselves of this definition, our model is still inadequate. For, with the 'nature' of every system (i.e. its levels ϵ_l) fixed once and for all, everything depends on one parameter, μ or U—or T. With a single variable the notion of 'integrating factor' collapses, for with dx, any $\phi(x)\,dx$ is also 'a complete differential'. Hence, to identify T, we must introduce the notion of other parameters or, what is the same thing, the notion of mechanical work done by the system.

Let us put, for the sake of brevity,

$$\log \sum_l e^{-\mu\epsilon_l} = F, \tag{2·11}$$

which is to be regarded as a function of μ and all the ϵ_l's; and let us write down, using (2·6), an undoubtedly correct mathematical relation, of which the physical application will follow presently:

$$\begin{aligned} dF &= \frac{\partial F}{\partial \mu} d\mu + \sum_l \frac{\partial F}{\partial \epsilon_l} d\epsilon_l \\ &= -U\,d\mu - \frac{\mu}{N} \sum_l a_l d\epsilon_l, \end{aligned} \tag{2·12}$$

and thus

$$d(F+U\mu) = \mu\left(dU - \frac{1}{N}\sum_l a_l d\epsilon_l\right). \tag{2·13}$$

We apply this to the following physical process, to which we subject our assembly of N systems.

We assume that each of them has identically the same 'mechanism' attached to it, screws, pistons and what not, which we can handle and thereby change its nature (i.e. the levels ϵ_l). We do so, changing, of course, the ϵ_l's for all of them alike in order that the basic condition of N identical systems, on which all our reasoning rests, shall be maintained. In addition, we also procure a direct 'change of temperature', by coupling our assembly with a large heat-bath (of the same temperature), changing the temperature of the whole very slightly and then isolating the assembly again from the heat-bath.

When (2·13) is applied to this process, $a_l d\epsilon_l$ is the work we have to do on the pistons, etc., attached to these a_l systems in order to 'lift them up' from the old level ϵ_l to the altered level $\epsilon_l + d\epsilon_l$; $\Sigma a_l d\epsilon_l$ is the work done in this way on the assembly, $-\Sigma a_l d\epsilon_l$ the work done by the assembly, and $-\frac{1}{N}\Sigma a_l d\epsilon_l$ the average work done by one of the members. And hence, since dU is its average energy increase, the round bracket to the right of (2·13) must be the average heat supply dQ supplied to it. μ is seen to be an integrating factor thereof. This alone really suffices to say that μ must be essentially $1/T$, because there is no other function of T which has this property for every system. And so $F + U\mu$ must be, essentially, the entropy.

To give a more direct proof, call

$$F + U\mu = G. \tag{2·14}$$

Then, from a general mathematical theorem, the ratio of the two integrating factors $1/T$ and μ is a function of G:

$$\frac{1}{T\mu} = \phi(G) \text{ say.} \tag{2·15}$$

Hence from (2·13)

$$\phi(G)\, dG = \frac{dQ}{T} = dS, \tag{2·16}$$

where S is the entropy. This, on integration, yields G as some function of S, say

$$G = \chi(S). \tag{2·17}$$

Now from (2·14), (2·11) and (2·6)

$$G = \log \sum_l e^{-\mu\epsilon_l} - \mu \frac{\partial}{\partial\mu} \log \sum_l e^{-\mu\epsilon_l}$$

'behaves additively' when two systems are combined (since $\log \Sigma e^{-\mu\epsilon_l}$ does). Calling the χ function of (2·17) χ_A in the case of a system A, χ_B for a system B and χ_{AB} for the combined system '$A+B$', and calling the entropies in these three cases S_A, S_B and S_{AB} respectively, we have

$$\chi_A(S_A)+\chi_B(S_B) = \chi_{AB}(S_{AB}).$$

On the other hand, the entropy, too, is an additive function, or at any rate

$$S_{AB} = S_A+S_B+C,$$

where C is independent of S_A and S_B. Hence

$$\chi_A(S_A)+\chi_B(S_B) = \chi_{AB}(S_A+S_B+C).$$

If you differentiate this equation once with respect to S_A, and again with respect to S_B, and compare the results, you get

$$\chi'_A(S_A) = \chi'_B(S_B) = \text{universal constant} = 1/k \text{ (say).}$$

Thus, from (2·17), (2·16) and (2·15)

$$\mu = \frac{1}{kT} \tag{2·18}$$

and $$S = kG+\text{const.} = k\log \sum_l e^{-\epsilon_l/kT} + \frac{U}{T}+\text{const.}, \tag{2·19}$$

where the const. is at any rate independent of T and of the 'parameters' (as volume, etc.) on which the ϵ_l's depend.

We drop the 'const.', pending an analysis of what that means. Then we have

$$k\log \sum_l e^{-\epsilon_l/kT} = S-\frac{U}{T} = \Psi. \tag{2·20}$$

We have thus obtained a general prescription—applicable to all cases (including the so-called 'new' statistics)—for obtaining the thermodynamics of a system from its mechanics.

Form the 'partition function' (also called 'sum-over-states'; German *Zustandssumme*)

$$Z = \sum_l e^{-\epsilon_l/kT}. \tag{2·21}$$

Then $k\log Z$ (where k is the Boltzmann constant) is the negative free energy, divided by T (with Planck we have introduced the letter Ψ for this function). From the comments made on (2·13) it is easily seen that our $k \log Z$ is the thermodynamical Ψ function in every respect, not only for changes of temperature but also for changes of the parameters (as, for example, the volume V) on which the ϵ_l may depend. Thus the average forces with which the system 'tends to increase these parameters' (e.g. the pressure p, in the case of the parameter volume) are found by the formulae familiar from thermodynamics, the prototype being

$$p = T\frac{\partial \Psi}{\partial V} = kT\frac{\partial}{\partial V}\log \sum_l e^{-\epsilon_l/kT}, \tag{2·22}$$

while the first equation (2·6),

$$U = -\frac{\partial}{\partial \mu}\log \sum_l e^{-\mu\epsilon_l} = T^2\frac{\partial \Psi}{\partial T}, \tag{2·23}$$

is an equally well-known formula of general thermodynamics (in all this Ψ is to be regarded as a function of T and such parameters as V, on which the ϵ_l may depend; macroscopically these parameters must fulfil the requirement, that when they are kept constant the system does no mechanical work*). Thus the statistical treatment, by yielding in principle Ψ as a function of T and the parameters like V, yields exhaustive information on the thermodynamical behaviour. (It is well known that a certain thermodynamical function yields complete information only when known as a function of certain variables. For example, $\Psi(T, p, \ldots)$ or $S(T, V, \ldots)$ does not, but, for example, $S(U, V, \ldots)$ does.)

* Against the treatment given here it could be objected that experimentally it is just as impracticable to keep such parameters as V exactly constant as to realize a complete heat isolation. A statistical treatment which introduces instead of the V's rather the quantities like p as parameters is perfectly possible, perhaps preferable, but considerably more complicated.

CHAPTER III

DISCUSSION OF THE NERNST THEOREM

WE turn to the question of what it means to have put the 'const.' zero in (2·19). Formally, it means adopting in every case a definite zero level for the entropy, which by elementary thermodynamics (excluding, for the moment, Nernst's theorem) is only defined by

$$dS = \frac{dQ}{T},$$

thus only up to an additive constant. Moreover, the zero level, thus adopted, is very simple and general. Indeed, if we write more explicitly

$$S = k \log \sum_l e^{-\epsilon_l/kT} + \frac{1}{T} \frac{\sum_l \epsilon_l e^{-\epsilon_l/kT}}{\sum_l e^{-\epsilon_l/kT}},$$

we can watch the behaviour of S at $T = 0$. Assuming for generality that the first n levels are equal ($\epsilon_1 = \epsilon_2 = \ldots = \epsilon_n$) and the following m levels ($\epsilon_{n+1} = \epsilon_{n+2} = \ldots = \epsilon_{n+m}$), then we can, for the purpose of finding the limit, certainly break off the sum after the $(n+m)$th term and obtain

$$S = k \log (ne^{-\epsilon_1/kT} + me^{-\epsilon_{n+1}/kT}) + \frac{1}{T} \frac{n\epsilon_1 e^{-\epsilon_1/kT} + m\epsilon_{n+1} e^{\epsilon_{n+1}/kT}}{ne^{-\epsilon_1/kT} + me^{-\epsilon_{n+1}/kT}}.$$

Considering that the second exponential becomes, near the limit, very small compared with the first, we easily get

$$S = k \log n - \frac{\epsilon_1}{T} + \frac{km}{n} e^{-(\epsilon_{n+1}-\epsilon_1)/kT} + \frac{\epsilon_1}{T} + \frac{1}{T}\frac{m}{n}(\epsilon_{n+1} - \epsilon_1) e^{-(\epsilon_{n+1}-\epsilon_1)/kT},$$

and thus

$$\lim_{T=0} S = k \log n.$$

This is practically zero, unless n were extremely large. To give an example: if the system were one mole of a gas (L molecules, say) and n were only of the order of L, $k \log L$ would be practically zero, because the order of magnitude that matters in this case is kL ($= \mathbf{R} =$ gas constant), against which $k \log L$ vanishes. But if we assumed that every molecule of the gas was capable of two different 'lowest states' with exactly the same energy, we should have $n = 2^L$, and $S(0) = kL \log 2 = \mathbf{R} \log 2$, which is appreciable. Modern gas theory assumes that such is not the case.

It is well known that to adopt $S = 0$ at $T = 0$ for every system is the conventional and most convenient way of pronouncing Nernst's famous heat theorem, sometimes called the Third Law. Have we then, by establishing (2·19) and by the subsequent considerations of this section, given the heat theorem a quantum-statistical foundation? At first sight it seems not, since our putting const. $= 0$ was an entirely arbitrary step.

Yet we have. For in point of fact—and contrary to what is often maintained—the numerical value of that const. is irrelevant, even meaningless. The relevant fact is that it is a constant, in other words, that that part of the entropy which does not vanish at $T = 0$ is independent of the 'parameters'. This fact entails the heat theorem statistically (as we shall immediately explain) in every particular case, and so in general, provided always that we exploit the idea of 'changing parameters' in the most general determination of which it is capable.

The mathematical part of the proof is simple enough: since the 'const.' is independent of the parameters, one and the same system approaches to the same entropy value, when you cool it down to zero, whatever the values of its parameters. In other words, the entropy difference of two thermodynamical states of the same system, differing by the values of the parameters, approaches to zero at $T = 0$.

Now the vanishing of this entropy difference is the only empirical content of the Nernst theorem. But in the truly important applications of the theorem, the two 'thermodynamical states' are so widely different that it needs a moment of reflexion to realize that they can be embraced by our notion: the same system with different values of the parameters.

A typical case would be a system consisting of L iron atoms* and L sulphur atoms. In one of the two thermodynamical states they form a compact body, 1 gram-molecule of FeS; in the other, 1 gram-atom of Fe and 1 gram-atom of S, separated by a diaphragm, so that they can under no circumstances unite; the much lower energy levels of the chemical compound are made inaccessible.

Now in all such cases it is only a question of believing in the possibility of transforming one state into the other by small reversible steps, so that the system never quits the state of thermodynamical equilibrium, to which all our considerations apply. All the small, slow steps of this process can then be regarded as small, slow changes of certain parameters, changing the values of the ϵ_l's. Then the 'const.' will not change in all these processes—and the statement applies.

For instance, in the example mentioned, you would gradually heat the gram-molecule of FeS till it evaporates; then go on heating till it dissociates as completely as desired; then separate the gases with the help of a semi-permeable diaphragm; then condense them separately by lowering the temperature (of course with an impermeable diaphragm between them) and cool them down to zero. Having once or twice gone through such considerations, you no longer bother to think them out in detail, but just declare them as 'thinkable'—and the statement applies.

After this has been thoroughly turned over in the mind the simplest way of codifying it once and for all is, of course, to decide to put 'const.' = zero in all cases. It is possibly the only way to avoid confusion—no alternative suggests itself. But to regard this 'putting equal to zero' as the essential thing is certainly apt to create confusion and to detract attention from the point really at issue.

* By L we mean 'Loschmidt's number', often called Avogadro's number, the number of molecules in one mole. We call it L because N is used up.

CHAPTER IV

EXAMPLES ON THE SECOND SECTION

FIRST a simple, but useful, remark. We have stated that

$$Z = \sum_l e^{-\mu\epsilon_l}$$

is 'multiplicative' and thus

$$\Psi = k \log Z = k \log \sum_l e^{-\mu\epsilon_l},$$

and all the other thermodynamical functions are strictly additive, if the system in question is made up of two or more systems in loose-energy contact, so that its *levels* ϵ_l are the sums of any level of the first system (α_k) and any level of the second system (β_m), etc.

A mathematically trivial, but physically relevant, remark is that the same 'multiplicative' or 'additive' composition obtains also if all the levels ϵ_l are the sums of two, or more, different kinds of levels in all combinations, even though the system itself is not really a juxtaposition of two, or more, systems.

For example, if the system is one gas molecule whose energy is the sum of its translational, rotational, and vibrational energy, all the thermodynamic functions are made up additively of a translational, a rotational, and a vibrational contribution—the mathematical situation being the same as if these three types of energy belonged to three independent systems in juxtaposition.

Similarly, we can deal with an ideal gas (L molecules in loose-energy contact) by first dealing with one molecule under the same conditions and then multiplying the thermodynamic functions by L. But that is, of course, nothing more than an application of the original idea of 'additivity' concerning two, or more, loosely coupled systems.

Very much more extended use of these remarks can be made than is done in the following simple examples.

(a) *Free mass-point* (ideal monatomic gas)

We are giving the old-fashioned, conventional treatment, dealing 'classically' with the mass-point. Without bothering about possible quantization we take as levels the cells of phase-space—the six-dimensional space of

$$x, \quad y, \quad z, \quad p_x, \quad p_y, \quad p_z.$$

The energy is $\frac{1}{2m}(p_x^2+p_y^2+p_z^2)$, m being the mass. Z is the *integral* (replacing here the *sum* over-states) of

$$e^{-\mu/2m}(p_x^2+p_y^2+p_z^2)\,dx\,dy\,dz\,dp_x\,dp_y\,dp_z$$

over the whole of phase-space (μ stands as an abbreviation for $1/kT$). Over the first three variables the integral is V, the volume, over the others it goes from $-\infty$ to $+\infty$. Thus

$$Z = V \iiint_{-\infty}^{+\infty} e^{-\mu/2m(p_x^2+p_y^2+p_z^2)}\,dp_x\,dp_y\,dp_z;$$

with an obvious transformation of variables

$$Z = V\left(\frac{2m}{\mu}\right)^{\frac{3}{2}} \iiint_{-\infty}^{+\infty} e^{-(\xi^2+\eta^2+\zeta^2)}\,d\xi\,d\eta\,d\zeta.$$

The integral is a constant, which does not interest us, since we are for the moment only interested in the derivatives of the logarithm of Z. Thus

$$\Psi = k\log Z = k\log V + \frac{3k}{2}\log T + \text{const.}$$

This for one atom. For L of them ($kL = \mathbf{R}$)

$$\Psi = \mathbf{R}\log V + \tfrac{3}{2}\mathbf{R}\log T + \text{const.}$$

From this we deduce, by (2·23) and (2·22),

$$U = T^2\frac{\partial\Psi}{\partial T} = \tfrac{3}{2}\mathbf{R}T, \quad p = T\frac{\mathbf{R}}{V},$$

the well-known formulae. This treatment is considered wrong

nowadays. The modern treatment will be given later. It involves the mathematical methods to be developed in the next section.

(b) Planck's oscillator

The levels are
$$\epsilon_l = (l+\tfrac{1}{2})h\nu \quad (l = 0,1,2,3,\ldots).$$

Hence
$$Z = \sum_{l=0}^{\infty} e^{-\mu h\nu(l+\frac{1}{2})} \quad \left(\text{put } \mu h\nu = \frac{h\nu}{kT} = x\right)$$
$$= e^{-\frac{1}{2}x}\sum_{l=0}^{\infty} e^{-lx} = e^{-\frac{1}{2}x}\frac{1}{1-e^{-x}} = \frac{1}{2\sinh\frac{1}{2}x}.$$

Hence
$$\Psi = k\log Z = -k\log\sinh\tfrac{1}{2}x - k\log 2,$$
$$U = T^2\frac{\partial\Psi}{\partial T} = -kT^2\frac{\cosh\frac{1}{2}x}{\sinh\frac{1}{2}x}\frac{1}{2}\left(-\frac{h\nu}{kT^2}\right)$$
$$= \frac{h\nu}{2}\frac{e^{\frac{1}{2}x}+e^{-\frac{1}{2}x}}{e^{\frac{1}{2}x}-e^{-\frac{1}{2}x}} = \frac{h\nu}{2}\frac{e^x+1}{e^x-1}$$
$$= \frac{h\nu}{2}+\frac{h\nu}{e^{h\nu/kT}-1},$$

which is the well-known expression, in which the 'zero-point energy' $\frac{1}{2}h\nu$ is usually dropped.

(c) Fermi oscillator

This is a particularly simple system (invented, as we shall see later, to formulate 'Fermi statistics'). It is a thing capable only of two levels, 0 and ϵ. Hence
$$Z = 1+e^{-\epsilon/kT},$$
$$\Psi = k\log(1+e^{-\epsilon/kT}),$$
$$U = T^2\frac{\partial\Psi}{\partial T} = kT^2\frac{e^{-\epsilon/kT}}{1+e^{-\epsilon/kT}}\frac{\epsilon}{kT^2}$$
$$= \frac{\epsilon}{e^{\epsilon/kT}+1}.$$

Compare this with the relevant second term on the right-hand side of the last equation of the preceding section (taking there

$\epsilon = h\nu$). There is just one remarkable difference in sign, ∓ 1 in the denominator. We shall see later that this constitutes the relevant difference between 'Einstein-Bose statistics' and 'Fermi-Dirac statistics'.

The thermodynamical functions of a system composed of L Planck oscillators or of L Fermi oscillators would, of course, be obtained on multiplying by L.

CHAPTER V

FLUCTUATIONS

To render the 'method of the most probable distribution', which we have used and which recommends itself by its great simplicity, entirely satisfactory, one would have to furnish a rigorous proof that its tacit assumption is justified, viz. that at least in the limit $N \to \infty$ (which we always mean when dealing with virtual Gibbs ensembles) the deviations from the 'most probable distribution' can be rigorously neglected.

It is worth while to mention one very plausible proof, though it is not quite flawless. (Quite good text-books offer it as a full proof, disguising it rather better than I propose to do here.)

Returning to the considerations (2·1), (2·2) and (2·3), we notice that the mean value of any a_m in all distributions is

$$\bar{a}_m = \frac{\Sigma a_m P}{\Sigma P}, \tag{5·1}$$

the sums to be understood over all sets a_l compatible with (2·3), while $a_m P$ in the numerator means that every P is to be multiplied by the particular value which the particular occupation number a_m has in that P.

Now change the definition of P formally by saying

$$P = \frac{N!}{a_1!\, a_2! \dots a_m! \dots} \omega_1^{a_1} \omega_2^{a_2} \dots \omega_m^{a_m} \dots, \tag{5·2}$$

on the understanding that the ω's have eventually to be equated to 1. Then (5·1) can be written

$$\bar{a}_m = \omega_m \frac{\partial \log \Sigma P}{\partial \omega_m} \tag{5·3}$$

(on the same understanding) and

$$\overline{a_m^2} = \frac{\Sigma a_m^2 P}{\Sigma P} = \frac{\omega_m}{\Sigma P} \frac{\partial}{\partial \omega_m}\left(\omega_m \frac{\partial \Sigma P}{\partial \omega_m}\right)$$

$$= \omega_m \frac{\partial}{\partial \omega_m}\left(\frac{\omega_m}{\Sigma P} \frac{\partial \Sigma P}{\partial \omega_m}\right) + \left(\frac{\omega_m}{\Sigma P} \frac{\partial \Sigma P}{\partial \omega_m}\right)^2.$$

Hence $$\overline{a_m^2}-(\bar{a}_m)^2 = \omega_m \frac{\partial}{\partial\omega_m}\left(\omega_m \frac{\partial \log \Sigma P}{\partial \omega_m}\right)$$
$$= \omega_m \frac{\partial \bar{a}_m}{\partial \omega_m}, \tag{5·4}$$

from (5·3). (Again all the ω_m's have to be equated to 1 eventually.)

All this is completely rigorous. But now we have to form an opinion about $\partial\bar{a}_m/\partial\omega_m$, i.e. of how $\bar{a}_m$ changes when ω_m changes. To do so, we have to interpret the preceding formula also with the ω's not equal to 1 (with at least ω_m differing slightly from 1). Now the expression (5·2) for P is actually often used in such considerations as that given in Chapter II, the ω's meaning the weights attributed to the various levels, according to their assumed degeneracy. Had we done so in Chapter II it would have made a very slight formal difference, namely, the $e^{-\mu\epsilon_l}$ would always be accompanied by ω_l, e.g. the 'most probable' a_l would be
$$a_m = N\frac{\omega_m e^{-\mu\epsilon_m}}{\sum\limits_l \omega_l e^{-\mu\epsilon_l}} \tag{5·5}$$
(to replace the second line in (2·6)).

From the preceding equation it is at least permissible to suggest that—with all the other $\omega_l = 1$ and only ω_m varying slightly in the neighbourhood of 1—a_m is very nearly proportional to ω_m. If that is admitted, then from (5·4) the view that in the limit $N \to \infty$ we have
$$\bar{a}_m = a_m$$
(the latter meaning the 'most probable') is consistent. For, indeed, then
$$\omega_m \frac{\partial \bar{a}_m}{\partial \omega_m} = \omega_m \frac{\partial a_m}{\partial \omega_m} = a_m = \bar{a}_m, \tag{5·6}$$
and
$$\overline{a_m^2}-(\bar{a}_m)^2 = \bar{a}_m. \tag{5·7}$$
That is to say, the 'dispersion' or fluctuation is 'normal' and vanishes practically, as N and thereby all the $\bar{a}_m$ go to infinity.

In most cases of application it is intuitively certain that the mean occupation number is very nearly proportional to the weight of the level, even for much larger changes of the weight of one level, e.g. if it is doubled or trebled. Indeed, with a big system it means a negligible modification of the system as a whole; and the two or three levels of the same description will, together, accommodate two or three times as many members of the ensemble. But that our conclusions are not entirely rigorous can be seen, if we use the first line of (5·4), thus:

$$\overline{a_m^2}-(\bar{a}_m)^2 = \omega_m \frac{\partial \log \Sigma P}{\partial \omega_m}+\omega_m^2 \frac{\partial^2 \log \Sigma P}{\partial \omega_m^2}$$

$$= \bar{a}_m = +\omega_m^2 \frac{\partial^2 \log \Sigma P}{\partial \omega_m^2}. \qquad (5{\cdot}8)$$

Here we see the term we have neglected. (It would be sufficient to prove, either that it is negative or that it is at most of the order of $\bar{a}_m$.)

An example of a system for which (5·7) fails—though a trivial one and one for which the dispersion is still smaller—is furnished by a single Fermi oscillator (forming the system—and, of course, N of them the Gibbs ensemble). We have in this case

$$P = \frac{N!}{a_0!\,a_1!}, \qquad (5{\cdot}9)$$

with $\qquad a_0+a_1 = N \quad \text{and} \quad 0.a_0+\epsilon.a_1 = E,$

hence $\qquad a_1 = \frac{E}{\epsilon} \quad \text{and} \quad a_0 = N-\frac{E}{\epsilon}.$

Hence the numbers are fixed, the dispersion is strictly zero. It is obvious that if we let the single system consist of two or four or five Fermi oscillators, the relation (5·7) would still not hold exactly, but hold only as to order of magnitude.

The 'method of mean values', explained in the following chapter, will yield an alternative proof of this order-of-magnitude-relation, i.e. of the vanishing of the dispersion or fluctuation in the limit $N \rightarrow \infty$.

Carefully to be distinguished from these (in the limit vanishing) fluctuations in the composition of the Gibbs ensemble are the fluctuations among the members of the ensemble, of which the ensemble is precisely the adequate representation, by containing systems in all sorts of different states $\epsilon_1, \epsilon_2, \ldots, \epsilon_l, \ldots$.

The simplest and a very important case is the fluctuation of the energy—the simple fact that the single systems have various energies, $\epsilon_1, \epsilon_2, \ldots, \epsilon_l, \ldots$, not all of them U. Now we had*

$$U = \frac{\Sigma \epsilon_l e^{-\mu \epsilon_l}}{\Sigma e^{-\mu \epsilon_l}} = \bar{\epsilon}_l \quad \left(\mu = \frac{1}{kT}\right).$$

Differentiate this with respect to μ (with the ϵ_l's constant, i.e. 'without external work'):

$$\frac{\partial U}{\partial \mu} = -\frac{\Sigma \epsilon_l^2 e^{-\mu \epsilon_l}}{\Sigma e^{-\mu \epsilon_l}} - \left(\frac{\Sigma \epsilon_l e^{-\mu \epsilon_l}}{\Sigma e^{-\mu \epsilon_l}}\right)^2,$$

giving

$$\overline{\epsilon_l^2} - (\bar{\epsilon}_l)^2 = -\frac{\partial U}{\partial \mu} = kT^2 \frac{\partial U}{\partial T}$$

or

$$\sqrt{\{\overline{\epsilon_l^2} - (\bar{\epsilon}_l)^2\}} = \sqrt{(kT \, . \, CT)}, \tag{5·10}$$

where we have introduced

$$C = \frac{\partial U}{\partial T}$$

for the heat capacity 'without external work'.

Equation (5·10) for the 'mean square fluctuation' has a very intuitive meaning. For many macroscopic systems at not too low a temperature CT can be regarded as roughly indicating the 'heat content', and this, grossly, as of the order nkT, where n is the number of degrees of freedom of the system. We see that in these cases the fluctuation is roughly of the order $kT\sqrt{n}$—which is very perspicuous to the statistician.

'Without external work' will as a rule mean 'with the parameters, as volume, kept constant'. I expressed it as I did in order to be able to include an interesting case with 'infinite' heat capacity and, therefore, 'infinite' fluctuations.

* The bar ($\bar{\epsilon}_l$) has now an entirely different meaning, which the reader will realize, without introducing a different notation.

If you enclose a fluid with its saturated vapour above it in a cylinder, closed by a piston, loaded with a weight to balance the vapour pressure—the piston gliding frictionlessly within the cylinder—and put it in a heat-bath, then you may include the piston and the weight in what you call the system and no 'external' work is done, even if the piston moves. Under these circumstances $C = \infty$, because any heat taken up or given off by the system will not change its temperature, but produce evaporation or condensation respectively. Any amount of fluctuation is thus to be expected, until either all the substance has been condensed or all evaporated.

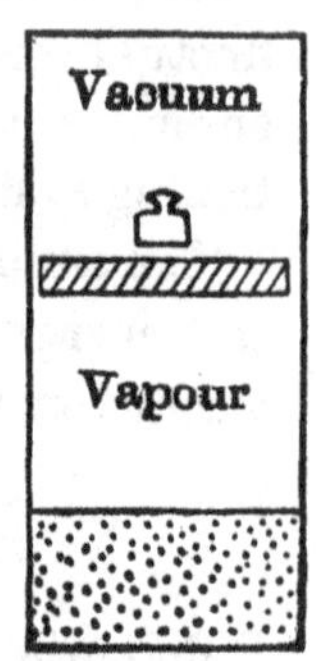

CHAPTER VI

THE METHOD OF MEAN VALUES

WE now resume the problem of Chapter II by a new method for several reasons. First, because the considerations of Chapter V failed to render the 'method of most probable values' entirely rigorous; the present method, which we owe to Darwin and Fowler, appeals to some scholars as being more convincing, perhaps even entirely exact. Secondly, it is always attractive and illuminating to see that identically the same result can be obtained by widely different considerations, especially if it is a question of a very general theorem of fundamental importance. Thirdly, the mathematical method to be developed here will prove very useful in other applications as well.

We aim at calculating actually the mean values of the a_l in the Gibbs ensemble, as indicated by (5·1). We avail ourselves of the manœuvre explained in (5·2), (5·3) and (5·4), in virtue of which all the desired information is reduced to the knowledge of the one quantity

$$\Sigma P = \sum_{(a_l)} \frac{N!}{a_1!\,a_2!\,\ldots\,a_l!\,\ldots}\,\omega_1^{a_1}\omega_2^{a_2}\ldots\omega_l^{a_l}\ldots, \tag{6·1}$$

the sum to be taken over all sets a_l that comply with (2·3). So all we have to do is to compute this sum.

If the only restriction on the a_l were $\Sigma a_l = N$, this task would be solved immediately by the polynomial formula and the sum would be

$$(\omega_1+\omega_2+\omega_3+\ldots+\omega_l+\ldots)^N,$$

at least formally (one would have to cut off the series of levels at some very high level to make the result finite). The second condition $\Sigma a_l \epsilon_l = E$ automatically restricts the number of terms in (6·1), because no level $\epsilon_l > E-(N-1)\,\epsilon_1$ can ensue, but at the same time it constitutes the real difficulty of the problem, which consists in selecting only the terms complying with this condition as well.

To cope with it we use the following artifice. We compute the following sum without the second restriction:

$$\Sigma P z^{a_1\epsilon_1+a_2\epsilon_2+\ldots+a_l\epsilon_l+\ldots}$$
$$= \Sigma \frac{N!}{a_1!\,a_2!\ldots a_l!\ldots}(\omega_1 z^{\epsilon_1})^{a_1}(\omega_2 z^{\epsilon_2})^{a_2}\ldots(\omega_l z^{\epsilon_l})^{a_l}\ldots$$
$$= (\omega_1 z^{\epsilon_1}+\omega_2 z^{\epsilon_2}+\ldots+\omega_l z^{\epsilon_l}+\ldots)^N = f(z)^N, \tag{6·2}$$

where
$$f(z) = \omega_1 z^{\epsilon_1}+\omega_2 z^{\epsilon_2}+\ldots+\omega_l z^{\epsilon_l}+\ldots. \tag{6·3}$$

Now supposing all the ϵ_l and E were integers, then the ΣP which we need is obviously the coefficient of z^E in the function (6·2) of z; it could be computed by the method of residues in the complex z plane.

To make this plan work, we must—and here the artifice comes in—declare that we have at the outset chosen the unit of energy so small that we can with any desired accuracy regard all the levels ϵ_l and the prescribed total energy E as integral multiples of this unit—or, if you prefer, replace them by integral multiples thereof without appreciably changing the physical problem. There are, of course, cases where this would appear to be impossible, in particular when the levels ϵ_l crowd infinitely dense near some finite energy ϵ, as is, for example, the case with the electronic levels in the hydrogen atom in open space. We exclude such cases, which, as can be shown, are altogether not amenable to any statistical treatment without special precautions (e.g. the hydrogen atom would have to be enclosed in a large but finite box, preventing the electron from escaping to infinity).

It is convenient to make two further restrictions about the ϵ_l. First, if $\epsilon_1 \neq 0$, we use the levels $0, \epsilon_2-\epsilon_1, \epsilon_3-\epsilon_1, \ldots, \epsilon_l-\epsilon_1, \ldots$ instead of $\epsilon_1, \epsilon_2, \ldots, \epsilon_l, \ldots$, replacing at the same time E by $E-N\epsilon_1$. A glance at (6·3) and at the following formula (6·4) shows that this makes no difference, it is only more convenient for the wording of our mathematical language. For the sake of simplicity we assume $\epsilon_1 = 0$. Secondly, we assume that the ϵ_l's have no common divisor. That can always be attained. For if they had, E would also have to have it, to make the condition $\Sigma a_l \epsilon_l = E$ strictly capable of fulfilment. Thus if τ be the greatest common divisor, we choose the energy unit τ times larger, which will remove the divisor, yet leave all the values integers.

Once this is agreed, the solution is simply and obviously

$$\Sigma P = \frac{1}{2\pi i}\oint z^{-E-1} f(z)^N \, dz, \tag{6·4}$$

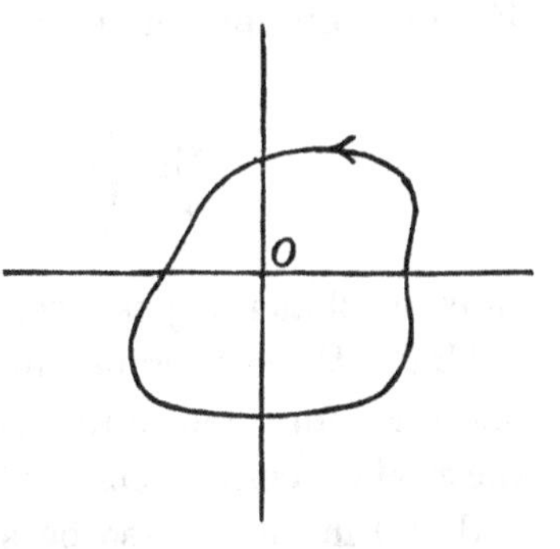

the integration to be conducted along any closed contour around the origin in the complex z plane—and, let me say, within the circle of convergence of $f(z)$, to avoid the need of analytical continuation.

The integral is evaluated by the method of steepest descent (German *Sattelpunktsmethode* = method of the saddle-point). Envisage the behaviour of the integrand, as you proceed from 0 to infinity on the real positive axis, remembering that in (6·3) all the ω's are virtually $= 1$ and that $0 = \epsilon_1 \leqslant \epsilon_2 \leqslant \epsilon_3 \ldots$. The first factor of the integrand, viz. z^{-E-1}, starts from an infinite positive value and decreases rapidly and monotonically. The second factor, viz. $f(z)^N$, starts at $z = 0$ from the value 1, increases monotonically, tending to infinity as z approaches the circle of convergence of $f(z)$, wherever that may be. Moreover, the relative decrease of the first factor, viz.

$$\frac{E+1}{z},$$

decreases itself monotonically from being $+\infty$ at $z = 0$ to 0 at $z = \infty$; the relative increase of the second factor, viz.

$$N\frac{f'(z)}{f(z)} = N\frac{\sum\limits_l \epsilon_l z^{\epsilon_l}}{\sum\limits_l z^{\epsilon_l}},$$

exhibits the opposite behaviour. It is zero at $z = 0$ and increases monotonically. Indeed,

$$\frac{d}{dz}\left(N\frac{f'(z)}{f(z)}\right) = N\frac{\sum\limits_{l}\epsilon_l^2 z^{\epsilon_l}\sum\limits_{k} z^{\epsilon_k} - (\sum\limits_{m}\epsilon_m z^{\epsilon_m})^2}{(\sum\limits_{s} z^{\epsilon_s})^2}.$$

Here the numerator can be written thus:

$$\sum_{l} z^{\epsilon_l}\left(\epsilon_l\sqrt{(\sum_{+\ k} z^{\epsilon_k})} - \frac{\sum\limits_{m}\epsilon_m z^{\epsilon_m}}{\sqrt{(\sum\limits_{+\ k} z^{\epsilon_k})}}\right)^2 > 0,$$

showing that it is positive.

Under these circumstances the integrand is bound to exhibit one and only one minimum (and no other extremum) within the circle of convergence of $f(z)$. This minimum may be expected, and will in due course be shown to be very steep, considering that both the exponents, viz. $E+1$ and N, are very large numbers. For, what we have actually and constantly in mind is the transition to the limits $N \to \infty$, $E \to \infty$, with the ratio E/N kept constant, since it is the average energy available to one system of the ensemble.

In other words, at this point on the real positive axis (which we shall call z_0 for the moment, but later drop the subscript zero again) the first derivative of the integrand vanishes and the second must be positive and can be anticipated to be very large. Hence if you proceed from this point orthogonal to the real axis, where the increment is purely imaginary, the integrand will exhibit (while remaining real at first) an exceedingly sharp maximum. We take for the contour of integration in (6·4) a circle, with the centre at O and passing through the point $z = z_0$, hoping that only the immediate neighbourhood of this very sharp maximum will essentially contribute to the value of the integral. We shall prove this in due course.

We first determine the value of z_0 by the vanishing of the first derivative and determine the value of the second derivative at

$z = z_0$. It is convenient to use logarithmic derivatives. Put, on the real positive axis,

$$z^{-E-1} f(z)^N = e^{g(z)} \tag{6·5}$$

(taking, of course, the main branch, i.e. the real value for $g(z)$). Then z_0 is determined by

$$g'(z_0) = -\frac{E+1}{z_0} + N\frac{f'(z_0)}{f(z_0)} = 0; \tag{6·6}$$

moreover
$$g''(z_0) = \frac{E+1}{z_0^2} + N\left(\frac{f''(z_0)}{f(z_0)} - \frac{f'(z_0)^2}{f(z_0)^2}\right). \tag{6·7}$$

This shows (i) that with E and N already very large, z_0 will not change, if you increase E and N proportionally; (ii) hence $g''(z_0)$ will in this case change proportionally with E and N and can thus be made, indeed, as large as you please (that it is positive need not be tested, since that follows from the general consideration).

Hence for a very small purely imaginary increment iy of z near $z = z_0$ the integrand can be written

$$z_0^{-E-1} (f(z_0))^N e^{-\frac{1}{2}y^2 g''(z_0) + -}, \tag{6·8}$$

and the neighbouring part of the circle of integration will (with any desired accuracy, if $g''(z_0)$ is made sufficiently large by increasing N) yield

$$[\Sigma P] = \frac{1}{2\pi i} z_0^{-E-1} f(z_0)^N \int_{-\infty}^{+\infty} e^{-\frac{1}{2}y^2 g''(z_0)}\, i\, dy$$

$$= z_0^{-E-1} f(z_0)^N \frac{1}{\sqrt{\{2\pi g''(z_0)\}}}. \tag{6·9}$$

We have enclosed ΣP in brackets, because it remains to be shown that this is all, that the contribution from the rest of the circle is negligible for N large.

* [The intuitive reason of this being so is, that the single terms of the series (6·3), which all 'reinforce' each other on the real axis, will, as z proceeds along the circle, 'rotate' round the

* The reader may interrupt the reading of the following lengthy proof, enclosed in [ ], wherever he pleases.

origin with different speeds, as the various integers ϵ_l prescribe; with the result that (outside the immediate neighbourhood of $z = z_0$, which has been taken care of) $|f(z)|$ will in general be considerably smaller than $f(z_0)$. Now the ratio of the absolute value of the integrand at an arbitrary point z of the circle to that at $z = z_0$ is

$$\left\{\frac{|f(z)|}{f(z_0)}\right\}^N, \tag{6·10}$$

which, for N large, will become arbitrarily small, also compared with the last factor in (6·9), viz. $(2\pi g''(z_0))^{-\frac{1}{2}}$, which is itself small, but only of the order $N^{-\frac{1}{2}}$. To make the conclusion rigorous, we have to show that the maximum value of $|f(z)|$, say M, is definitely smaller than $f(z_0)$:

$$M < f(z_0). \tag{6·11}$$

For, then, the contribution of the rest of the circle to the absolute value of the integral is certainly not larger than

$$\frac{1}{2\pi} z_0^{-E} M^N 2\pi = z_0^{-E} f(z_0)^N \left(\frac{M}{f(z_0)}\right)^N, \tag{6·12}$$

which, for $N \to \infty$, is negligible compared with (6·9).

To prove (6·11) we observe that equality, $M = f(z_0)$, could only occur if at some point z on the circle, definitely different from z_0, all the terms in (6·3) again reinforced each other. Since the first term is real and positive ($\epsilon_1 = 0$), they would all have to be real and positive there. Let ϕ ($\leqslant 2\pi$) be the phase angle at that point. Then all the products

$$\epsilon_1\phi, \quad \epsilon_2\phi, \quad \ldots, \quad \epsilon_l\phi, \quad \ldots,$$

would have to be integral multiples of 2π and all the integers ϵ_l integral multiples of $2\pi/\phi$, say

$$\epsilon_l = n_l \frac{2\pi}{\phi}.$$

But this cannot be, unless $\phi = 2\pi$ (i.e. at $z = z_0$). For, if $2\pi/\phi$ were >1, it would have to be a rational fraction p/q with a numerator larger than 1, even when written with smallest integrals p, q. Then p would be a common divisor of all the ϵ_l, which is contrary to our assumption that there should be none.

This proof is rather sophisticated and not very satisfactory to the physicist, who hesitates to believe that one single level ϵ_l could all but upset the apple-cart. Indeed, what might conceivably happen is that all but one have a fairly large common divisor p, which could not be removed on account of the one which does not possess it. It is therefore well to be satisfied that even such a 'single dissenter' would prevent the maximum M from approaching arbitrarily near to $f(z_0)$. Indeed, since not all the ϵ_l are to have a common divisor, they must attain this property (the property, that is, of having none) at some finite point of the series, say ϵ_m. The supposed 'dissenter' can then only occur for $\epsilon_l \leqslant \epsilon_m$, and that obviously also sets an upper limit to the supposed common divisor p of the rest. The not-wholly-real term of the series would in this case have at least the phase angle $2\pi/p$ and would then read

$$\omega_l z_0^{\epsilon_l} e^{2\pi i/p}.$$

This obviously produces a finite departure of $|f(z)|$ from $|f(z_0)|$, though, with ϵ_l and p fairly large, the departure might be fairly small; the rest must be taken care of by passing to the limit $N \to \infty$ in (6·10) or (6·12).]

Let us now return to our essential results (6·6), (6·7) and (6·9). In rewriting them we drop for brevity the index in z_0, because no other than this one real positive value of z concerns us, and we also understand z in (6·3) to mean this value. So, collecting our results, we have

$$f(z) = \omega_1 z^{\epsilon_1} + \omega_2 z^{\epsilon_2} + \ldots + \omega_l z^{\epsilon_l} + \ldots, \qquad (6\cdot13)$$

$$g'(z) = -\frac{E+1}{z} + N\frac{f'(z)}{f(z)} = 0, \qquad (6\cdot14)$$

$$g''(z) = \frac{E+1}{z^2} + N\left(\frac{f''(z)}{f(z)} - \frac{f'(z)^2}{f(z)^2}\right), \qquad (6\cdot15)$$

$$\Sigma P = z^{-E-1} f(z)^N \frac{1}{\sqrt{\{2\pi g''(z)\}}}, \qquad (6\cdot16)$$

$$\log \Sigma P = -(E+1)\log z + N \log f(z) - \tfrac{1}{2}\log(2\pi g''(z)). \qquad (6\cdot17)$$

Even the last term in the last formula will turn out to be negligible, and we might drop it forthwith on the ground that it is only of the order of $\log N$. But to be on the safe side we still keep it by for a while.

From (5·3) we now obtain the mean occupation numbers

$$\bar{a}_l = \omega_l \frac{\partial \log \Sigma P}{\partial \omega_l} = \omega_l g'(z) \frac{\partial z}{\partial \omega_l} + \frac{\omega_l N z^{\epsilon_l}}{f(z)} - \frac{1}{2} \frac{\partial}{\partial \omega_l} \log g''(z). \quad (6·18)$$

The first term is zero by (6·14). (But, of course, we were obliged to take the implicit dependence of z on ω_l into account.) As regards the last term, let us introduce the average energy

$$\frac{E}{N} = U, \quad (6·19)$$

which does not change in the limiting process $N \to \infty$, $E \to \infty$. Then (6·15) reads

$$g''(z) = N\left(\frac{U}{z^2} + \frac{f''}{f} - \frac{f'^2}{f^2}\right). \quad (6·20)$$

Hence the last term in (6·18) is also constant in the limiting process and we have (putting all the $\omega_l = 1$, according to plan)

$$\bar{a}_l = N \frac{z^{\epsilon_l}}{z^{\epsilon_1} + z^{\epsilon_2} + \ldots + z^{\epsilon_l}}. \quad (6·21)$$

The equation (6·14), which determines z, can be written, with (6·19) and putting all the ω's equal to 1,

$$U = \frac{\epsilon_1 z^{\epsilon_1} + \epsilon_2 z^{\epsilon_2} + \ldots + \epsilon_l z^{\epsilon_l} + \ldots}{z^{\epsilon_1} + z^{\epsilon_2} + \ldots + z^{\epsilon_l} + \ldots}. \quad (6·22)$$

The last two equations, if we put

$$\log z = -\mu, \quad (6·23)$$

are an identical replica of the fundamental relations (2·6), on which we built the thermodynamical theory from that point onwards. Only, the mean values $\bar{a}_l$ now have replaced the most probable values. Our $f(z)$ plays the part of the 'sum-over-states'. So, we may now claim that we have founded the theory in a second, independent manner. Let us see what we now get for the fluctuation of $\bar{a}_l$, from (5·4). Using (6·18) we have to form

$$\overline{a_l^2}-(\bar{a}_l)^2 = \omega_l \frac{\partial}{\partial \omega_l}\left\{\omega_l g'(z)\frac{\partial z}{\partial \omega_l}+\frac{\omega_l N z^{\epsilon_l}}{f(z)}-\frac{1}{2}\frac{\partial}{\partial \omega_l}\log g''(z)\right\}.$$

Here the first term gives nothing, because $g'(z) = 0$ (6·14) must be understood to hold identically in ω_l. The last term can be dropped, because, from (6·20), it has 'no order' in N—and terms of the order N will survive. In differentiating the middle term, we must again take into account that z depends on ω_l (though, as a rule, not much, and, from (6·14) and (6·19), in a way that does not depend on N). We get

$$\overline{a_l^2}-(\bar{a}_l)^2 = \omega_l N\frac{z^{\epsilon_l}}{f(z)}+\omega_l^2 N\left\{\frac{\partial z}{\partial \omega_l}\left(-\frac{z^{\epsilon_l}f'(z)}{f(z)^2}+\frac{\epsilon_l z^{\epsilon_l-1}}{f(z)}\right)-\frac{z^{2\epsilon_l}}{f(z)^2}\right\}.$$

Putting all the ω's equal to 1 and using (6·14), (6·19) and (6·21) we easily obtain

$$\overline{a_l^2}-(\bar{a}_l)^2 = \bar{a}_l\left[1+(\epsilon_l-U)\frac{\partial \log z}{\partial \omega_l}-\frac{\bar{a}_l}{N}\right]. \qquad (6·24)$$

Since the square bracket certainly does not involve the order N, the mean square deviation is, if not precisely 'normal', certainly 'of normal order', i.e. of the order of $\bar{a}_l$. Thus the relative fluctuation approaches to zero, as N and all the $\bar{a}_l$ approach to infinity. The distribution becomes infinitely sharp. Mean values, most probable values, any values that occur with non-vanishing probability—all become the same thing.

But it is possible and quite illuminating (though not very important) to evaluate the middle term in (6·24) exactly. It turns out to be always negative, so that the fluctuations are infra-normal. For that purpose it is slightly more convenient to liquidate the z notation and to replace it by the μ- or T notation, according to (6·23):

$$\log z = -\mu = -\frac{1}{kT}.$$

Thus

$$\frac{\partial \log z}{\partial \omega_l} = -\frac{\partial \mu}{\partial \omega_l}\left(=\frac{1}{kT}\frac{\partial \log T}{\partial \omega_l}\right).$$

This dependence of μ on one ω_l is to be calculated from (6·14), which can be written

$$U = \frac{\Sigma \epsilon_l \omega_l e^{-\mu\epsilon_l}}{\Sigma \omega_l e^{-\mu\epsilon_l}}.$$

The understanding is that $U = \text{const.}$ Thus

$$d \log U = \frac{ds_1}{s_1} - \frac{ds_0}{s_0} = 0,$$

where we put for brevity

$$s_k = \Sigma \epsilon_l^k \omega_l e^{-\mu\epsilon_l}.$$

Now (varying only μ and one ω_l),

$$ds_1 = \epsilon_l e^{-\mu\epsilon_l} d\omega_l - s_2 d\mu, \quad ds_0 = e^{-\mu\epsilon_l} d\omega_l - s_1 d\mu.$$

Hence $$\frac{\epsilon_l e^{-\mu\epsilon_l} d\omega_l - s_2 d\mu}{s_1} - \frac{e^{-\mu\epsilon_l} d\omega - s_1 d\mu}{s_0} = 0.$$

From this we easily obtain

$$\frac{\partial \mu}{\partial \omega_l} = \frac{\epsilon_l - s_1/s_0}{s_2/s_0 - s_1^2/s_0^2} \frac{e^{-\mu\epsilon_l}}{s_0}.$$

Considering the meaning of the symbols, that reads*

$$\frac{\partial \mu}{\partial \omega_l} = \frac{\epsilon_l - U}{\widetilde{(\epsilon_l - U)^2}} \frac{\bar{a}_l}{N}.$$

And so we get from (6·24)

$$\overline{a_l^2} - (\bar{a}_l)^2 = \bar{a}_l \left[1 - \frac{(\epsilon_l - U)^2}{\widetilde{(\epsilon_l - U)^2}} \frac{\bar{a}_l}{N} - \frac{\bar{a}_l}{N} \right].$$

We should call the dispersion 'normal' if the middle term were zero; and so it is for those levels which amount to the mean energy ($\epsilon_l = U$). In all other cases the dispersion is infra-normal.†

One of the fascinating features of statistical thermodynamics is that quantities and functions, introduced primarily as mathe-

* The wavy line $\sim$ indicates the mean value among the members of the assembly, as contemplated towards the end of Chapter v.

† I remind you of the case of a single Fermi oscillator, where the fluctuation of $\bar{a}_l$ proved to be strictly zero.

matical devices, almost invariably acquire a fundamental physical meaning. We have had examples in the Lagrangian parameter μ, the maximum z, the sum-over-states or partition function. What is the meaning of ΣP? We take it from (6·17), drop the last term on account of its smallness, and use the notation (6·19) as well as (6·23), remembering that $\mu = 1/kT$. Then

$$\frac{1}{N}\log \Sigma P = \frac{U}{kT} + \log f(z)$$

$$= \frac{U}{kT} + \frac{1}{k}\left(S - \frac{U}{T}\right) = \frac{1}{k}S.$$

Thus
$$\frac{k}{N}\log \Sigma P = S, \tag{6·25}$$

the entropy of the single system. This is quite interesting in itself, but becomes more remarkable if we return for a moment to Chapter II and compute the logarithm of the maximum P (which we could have done already there, but we did not). From (2·2) and Stirling's formula

$$\log P = N(\log N - 1) - \sum_l a_l(\log a_l - 1)$$

$$= N \log N - \sum_l a_l \log a_l.$$

Using the 'maximum' a_l's

$$a_l = N\frac{e^{-\mu\epsilon_l}}{\sum\limits_l e^{-\mu\epsilon_l}},$$

and thus
$$\log a_l = \log N - \mu\epsilon_l - \log \sum_l e^{-\mu\epsilon_l},$$

we obtain

$$\log P_{\text{max.}} = N \log N - \sum_l a_l \log N + \mu \Sigma a_l\epsilon_l + \Sigma a_l \log \Sigma e^{-\mu\epsilon_l}$$

$$= \mu E + N \log \Sigma e^{-\mu\epsilon_l},$$

and further
$$= \frac{E}{kT} + N\frac{1}{k}\left(S - \frac{U}{T}\right).$$

Multiply by k/N and remember that $E/N = U$:

$$\frac{k}{N}\log P_{\text{max.}} = \frac{U}{T} + S - \frac{U}{T} = S. \tag{6·26}$$

Comparing this with (6·25), we find that we can calculate the entropy either as $\log \Sigma P$, or as $\log P_{\text{max.}}$; it makes no difference. What happens is that the number of P's that are quite comparable with $P_{\text{max.}}$ is very large, yet vanishingly small in comparison with $P_{\text{max.}}$ itself. Hence in the logarithms the distinction is negligible. This is what H. A. Lorentz called in a famous mémoire: 'L'insensibilité des fonctions thermodynamiques.'

There are other statistical analogues of the entropy, but not of such general applicability as this. This, derived from the sum-over-states, is applicable to any system, whether large or small, to a single oscillator as well as to a gas, a solid or a system of several phases.

The one now to be indicated (due also to W. Gibbs) presupposes a system showing only small energy fluctuation in a heat-bath, as we know to be the case with any big system. Only the levels very near to the average energy U are occupied. But look at the sum-over-states

$$e^{-\mu\epsilon_1}+e^{-\mu\epsilon_2}+\ldots+e^{-\mu\epsilon_l}+\ldots.$$

Since the ϵ's are in arithmetical order, the exponentials decrease permanently. Now, they are a measure of occupation frequency! At first sight we are astonished how that sharp maximum should come about—why any maximum at all?

The cause lies in the way in which ϵ_l increases with l, namely, slower and slower as you proceed in the series, and indeed with 'tremendously increasing slowness'. In other words, the number of levels per unity increase of ϵ_l, the level density, increases enormously. The maximum comes about as a compromise between the increasing level density and the decreasing exponential.

Let us look at it in this way: we may regard ϵ_l as a function of its subscript, $\epsilon(l)$; and thus also l as a function of ϵ, $l(\epsilon)$—in words: the number of levels up to the limit ϵ. Now take

$$l(U),$$

where U is the average energy (from which actually only small deviations occur at all). Then

$$k \log l(U)$$

is the entropy.

This is not difficult to grasp; but, in the first instance, we shall come upon yet another 'entropy definition'.

Let us choose some convenient small step $\Delta\epsilon$ and bundle together all the levels within such a $\Delta\epsilon$ step, say Δl. Then the sum-over-states can be written

$$\Sigma e^{-\mu\epsilon} \Delta l,$$

ϵ being, of course, the value within that region Δl. Then we may also write

$$\Sigma e^{-\mu\epsilon} \frac{\Delta l}{\Delta\epsilon} \Delta\epsilon.$$

The region of maximum occupation—and that is the region $\epsilon \sim U$—is determined by the maximum of the 'integrand', or, if you like, of its logarithm

$$-\epsilon\mu + \log \frac{\Delta l}{\Delta\epsilon}.$$

Thus
$$-\mu + \left(\frac{d}{d\epsilon} \log \frac{\Delta l}{\Delta\epsilon}\right)_{\epsilon=U} = 0,$$

$$\frac{1}{T} = \left(k \frac{d}{d\epsilon} \log \frac{\Delta l}{\Delta\epsilon}\right)_{\epsilon=U},$$

$$\frac{1}{T} = \left(\frac{dk \log \dfrac{\Delta l}{\Delta\epsilon}}{d\epsilon}\right)_{\epsilon=U}.$$

This shows, of course, that

$$S = \left(k \log \frac{\Delta l}{\Delta\epsilon}\right)_{\epsilon=U}$$

plays the role of entropy.

The reason why we may take $l(U)$ itself instead of

$$\frac{\Delta l}{\Delta\epsilon} \quad \text{or} \quad \frac{dl(U)}{dU}$$

is that $l(U)$ increases practically always as some enormously high power of U,

$$l(U) = CU^n \quad \text{(say)},$$

$$\frac{dl}{dU} = nCU^{n-1},$$

$$\log l = \log C + n \log U,$$

$$\log \frac{dl}{dU} = \log(nC) + (n-1)\log U.$$

You see, it makes practically no difference.

I should like to indicate the intuitive reason for the exponential dependence of the occupation frequency on ϵ, in heat-bath conditions.

Let $\epsilon_1, \epsilon_2, \ldots, \epsilon_l$, etc., be the energy levels of the system and $b_1, b_2, \ldots, b_k$, etc., be the levels of the bath. Then the sum of the total energy (E) of the system plus that of the bath is a constant, and the levels of the whole are $\epsilon_l + b_k$.

Since the total energy is constant, there is actually only an exchange between degenerate levels, i.e.

$$\epsilon_l + b_k = \text{(nearly) constant} = E.$$

('Nearly' on account of the coupling energy!) Now all these levels $\epsilon_l + b_k$ for all combinations (l, k) have, of course, equal occupation frequency, that is, simply the assumption of equal *a priori* probability for any single level. The reason for the decreasing occupation frequency of the higher ϵ_l is that the number of bath levels

$$b_k = E - \epsilon_l$$

decreases exponentially with decreasing $E - \epsilon_l$, the energy left to the bath. That is, indeed, pretty clear if this number is something like

$$C(E-\epsilon_l)^n = CE^n\left(1-\frac{\epsilon_l}{E}\right)^n \approx CE^n e^{-n\epsilon_l/E}.$$

$$(n = \text{very large})$$

This does not pretend to be a strict derivation (we have given that before), but it shows the gist of the thing: the more of the total energy (E) the system assumes for itself (ϵ_l), the less (viz. $E-\epsilon_l$) is left to the bath. And this reduces the number of available bath-levels—even in the case of an infinite heat-bath, or rather, precisely in this case—by an exponential factor, the very one which we have come to know as the corresponding term in the sum-over-states, that is to say, as the relative probability of finding our system in a state ϵ_l under heat-bath conditions.

Chapter VII

THE n-PARTICLE PROBLEM

So from now on we need no longer distinguish (for the purpose of dealing with a Gibbs ensemble, $N \to \infty$) between the two methods, that of the 'most probable values' and that of 'mean values'. For they have produced exactly the same result. In fact, since this result, once known, is quite generally applicable to every system, we need never recur to either of the methods! Looking back, we might indeed have scrapped the first altogether, which is, so to speak, legalized only by the second. The only reason why I did not scrap the first was, that its mathematics—consisting really only of two or three lines—is so much more easily surveyable; and that is worth a lot in a domain which is conceptionally so enormously difficult.

But though I have just said that we shall never more have to return to either method, the mathematical idea of Darwin-Fowler, which overcomes the difficulty of an accessory condition by forming a residue, is such a sublimely excellent device that we shall indeed take to it again, namely, for the purpose of evaluating the sum-over-states in cases which otherwise would be clumsy to handle. But I beg you to keep those two things apart in your mind: the general proof is done with. When we use complex integration in what follows, it is by way not of giving an example of the general method, but of using a similar mathematical device or tool for evaluating certain sums-over-states. It is essential to emphasize this point. For when, having first explained a general method, one proceeds to use very much the same mathematical device in dealing with particular examples, the inference is almost bound to be that he has applied the general method to the special example!

Now for the n-particle problem (simplest application: an ideal gas).

According to modern views, a gas must not be regarded as consisting of n identical systems in loose-energy contact, since the energy levels of the gas are not the sums of the energy levels of its n constituents in all combinations. They are numerically equal to them, certainly.* But any two gas levels which differ only by an exchange of roles between two (or more) of the n identical atoms or molecules, have to be regarded as one and the same level of the gas. Brief reflexion will show that this produces an entirely different sum-over-states for the gas as a whole.

The underlying physical idea is that the particles are energy quanta without individuality; that Democritos of Abdera, not Max Planck, was the first quantum physicist. For the moment, we defer discussion of the physical meaning and of the experimental facts which have forced this entirely new attitude upon us. We first devote ourselves to determine, from our general theory, the new thermodynamics of the n-particle system. Denoting by

$$\alpha_1, \quad \alpha_2, \quad \alpha_3, \quad \ldots, \quad \alpha_s, \quad \ldots,$$

the levels of one particle, a definite level (not a class of levels!) ϵ_l of the n-particle system (we will call it 'the gas', for short) is indicated by the numbers

$$n_1, \quad n_2, \quad n_3, \quad \ldots, \quad n_s, \quad \ldots,$$

of particles on level $\alpha_1, \alpha_2, \ldots$ respectively; and that level ϵ_l is

$$\epsilon_l = n_1\alpha_1 + n_2\alpha_2 + \ldots + \ldots = \Sigma n_s\alpha_s.$$

(Not to be confused with a previous scheme—the a_l's and the ϵ_l's—to which it bears only a formal resemblance.)

Hence the sum-over-states is ($\mu = 1/kT$)

$$Z = \sum_{(n_s)} e^{-\mu \Sigma n_s \alpha_s}. \tag{7·1}$$

The $\sum\limits_{(n_s)}$ means: over all admissible sets of numbers n_s. This expression embraces several different physical cases: the theory

* In the case of no interaction between the particles. We envisage only this case in these lectures.

of black-body radiation; the theory of ordinary Bose-Einstein gases, and thereby the theory of the so-called chemical constant; the theory of a Fermi-Dirac gas, of which the most important application is to the electrons in metals. We shall evaluate Z in all these cases. From $\log Z$ we then deduce the thermodynamics. Another point of interest is the average value of n_s. Let us note, for future use, that it is always

$$\bar{n}_s = -\frac{1}{\mu}\frac{\partial \log Z}{\partial \alpha_s}, \tag{7·2}$$

as may be verified at a glance. But, mind you, if the system is finite, the fluctuations of these n_s are not entirely negligible. The case is quite different from the fluctuations of the a_l in a virtual assembly.

The different cases in the evaluation of Z arise thus:

(i) The values admitted for every n_s may be

(*a*) $n_s = 0, 1, 2, 3, 4, \ldots$ (Bose-Einstein gas).

(*b*) $n_s = 0, 1$ (Fermi-Dirac gas; Pauli's exclusion principle).

(ii) There may or may not be the condition that the total number of particles is constant,

$$\sum_s n_s = n. \tag{7·3}$$

Only one case is known, though, where this condition is not imposed; it is a Bose-Einstein case, viz. black-body radiation (photons). It is, of course, the simplest one.

Put in any case $$z_s = e^{-\mu\alpha_s}; \tag{7·4}$$

thus $$Z = \sum_{(n_s)} z_1^{n_1} z_2^{n_2} \ldots z_s^{n_s} \ldots, \tag{7·5}$$

to be summed over all admissible sets of numbers n_s.

Paying attention, at first, only to the restrictions (i) (*a*) or (i) (*b*) respectively, we easily obtain

(i) (*a*) $$Z = \prod_s \frac{1}{1-z_s} \quad \text{(Bose-Einstein)},$$

(*b*) $$Z = \prod (1+z_s) \quad \text{(Fermi-Dirac)}.$$

It is convenient to combine these formulae thus:

$$Z = \prod_s (1 \mp z_s)^{\mp 1}, \tag{7·6}$$

where the double signs are coupled and the upper sign refers throughout to the Bose-Einstein case.

We have still disregarded (7·3). As I have already said, this is the correct attitude only in one particular case (heat radiation; upper sign). We might follow up this simplest case first. It is suggestive, but it would not be economical to do so.

When (7·3) is imposed, (7·6) is not yet the final result. For a glance at the original form (7·5) indicates that we have to select from (7·6) only the terms homogeneous of order n in all the z_s. That is most conveniently done by the method of the residue. Put

$$f(\zeta) = \prod_s (1 \mp \zeta z_s)^{\mp 1}. \tag{7·7}$$

Then the correct Z is rigorously represented by the following integral:

$$Z = \frac{1}{2\pi i} \oint \zeta^{-n-1} f(\zeta)\, d\zeta, \tag{7·8}$$

conducted around the origin in the complex ζ plane in such a way that no other singularity of the integrand, except $\zeta = 0$, is embraced.

It is not very difficult to show that in both cases the integrand on the real positive axis starts at $\zeta = 0$ from large positive values, while its logarithmic derivative starts from large negative values and, increasing continually, becomes eventually positive. Hence the integrand has one and only one minimum on the way and the method of steepest descent can be tried.

Putting, on the real positive axis,

$$\zeta^{-n-1} f(\zeta) = e^{g(\zeta)}, \tag{7·9}$$

we obtain the following two relations:

$$g'(\zeta) = -\frac{n+1}{\zeta} + \frac{f'(\zeta)}{f(\zeta)} = 0, \tag{7·10}$$

$$g''(\zeta) = \frac{n+1}{\zeta^2} + \frac{f''(\zeta)}{f(\zeta)} - \frac{f'(\zeta)^2}{f(\zeta)^2}, \tag{7·11}$$

the first of which determines a real positive root ζ (we omit the embellishment of a subscript zero), while the second indicates the value of $g''(\zeta)$ at that point. And we get

$$Z = \zeta^{-n-1} f(\zeta) \frac{1}{\sqrt{\{2\pi g''(\zeta)\}}}, \tag{7·12}$$

$$\log Z = -(n+1)\log\zeta + \log f(\zeta) - \tfrac{1}{2}\log(2\pi g''(\zeta)), \tag{7·13}$$

pending the proof that $g''(\zeta)$ is very large also in the present case.

Though in the present case we cannot strictly pass to the limit $n \to \infty$, we may virtually do so. To begin with, we safely replace in (7·10) $n+1$ by n and obtain, with (7·7) and (7·9),

$$n = \Sigma \frac{1}{\frac{1}{\zeta} e^{\mu\alpha_s} \mp 1} \quad \left(\mu = \frac{1}{kT}\right). \tag{7·14}$$

We shall presently replace this sum by an integral, whereby it will turn out to be proportional to the volume V, by virtue of the fact that, for fairly large n, the number of levels α_s between any given narrow energy limits is proportional to V. Hence the characteristic root ζ only depends on the volume density of particles n/V. That is not rigorously true for any finite n, but we just declare that we wish to investigate only the limiting behaviour of sufficiently large 'gas bodies'. ζ being thus fixed, (7·11) shows that $g''(\zeta)$ is actually arbitrarily large, if n is. Hence not only the procedure that led to (7·12) and (7·13) is justified, but also the neglect of the last term in (7·13), because it is only of the order of $\log n$; thus

$$\log Z = -n\log\zeta + \log f(\zeta) = -n\log\zeta \mp \sum_s \log(1 \mp \zeta e^{-\mu\alpha_s}). \tag{7·15}$$

(To refuse the virtual transition to the limit $n \to \infty$ would lead us to something we are not interested in for the moment, namely, gas bodies so small that their thermodynamical behaviour depends on their size and shape. The peculiar features that would result would be termed 'surface phenomena' by the experimentalist.) The parameter ζ in (7·15) is determined by (7·10), which is more easily surveyed in the form (7·14). On the other hand, we can now see that it was a time-saving device, not to treat separately the case in which the total number of particles is not prescribed but is, so to speak, allowed to adjust itself. In this case Z was given directly by (7·6), with (7·4). It is easily seen that this case is formally embraced by our present value of $\log Z$, viz. by (7·15), if we just put $\zeta = 1$ (instead of letting it be determined by (7·14); yet the latter equation is not entirely meaningless even now; it gives us the changing number n of particles actually present).

The partial derivative of (7·15) with respect to ζ vanishes, according to (7·10). Hence (though ζ, by (7·10), does depend on α_s) we get from (7·2)

$$n_s = -\frac{1}{\mu}\frac{\partial \log Z}{\partial \alpha_s} = \frac{1}{\frac{1}{\zeta}e^{\mu\alpha_s} \mp 1}, \tag{7·16}$$

which renders the equation (7·14) very translucent.

From the meaning of the last equation the average energy U of our gas body is obviously

$$U = \sum_s \frac{\alpha_s}{\frac{1}{\zeta}e^{\mu\alpha_s} \mp 1}. \tag{7·17}$$

I beg the reader to verify for himself that this could also have been obtained from (7·15) by the general relation

$$U = T^2\frac{\partial \psi}{\partial T} = kT^2\frac{\partial \log Z}{\partial T}. \tag{7·18}$$

With one thrilling exception (the case of Bose-Einstein condensation, which we shall discuss in detail later) the sums in (7·14), (7·15) and (7·17) and similar ones can be evaluated as integrals, without our having to bother about the exact values of the levels α_s; we are concerned only with their density per unit of energy increase.

We restrict ourselves to the case that α_s represents only translatory energy. (At the low temperatures where the 'new' statistics differ from the 'old', every gas has become 'monatomic', vibrations and molecular rotations having died down entirely. So there would be no point in including them.)

The number of states (of the single particle) pertaining to a 'physically infinitesimal' element of phase-space is

$$\frac{dx\,dy\,dz\,dp_x\,dp_y\,dp_z}{h^3}.$$

Integrating with respect to the first three variables over the volume V and also over the '4π directions' of the momentum, we have

$$\frac{4\pi V}{h^3}p^2dp, \tag{7·19}$$

where p is the absolute value of the momentum. If the particles are endowed with spin, this number has still to be multiplied by a small integer, 2 or 3, according to the different orientations of the spin that are possible (2 for spin $\frac{1}{2}$ and also for spin 1, when the rest-mass vanishes (photon); 3 for spin 1, when the rest-mass does not vanish (meson)).

(7·19) is the distribution of single-particle states 'on the momentum line' p. What we need for evaluating our sums is the distribution on the energy line 'α'. The general relation between α and p for a free particle is, of course,

$$\alpha = c\sqrt{(m^2c^2+p^2)}. \tag{7·20}$$

It would embrace all cases. But the square root makes it inconvenient. It can be avoided, since in point of fact only the two limiting cases actually arise, viz.

either (i) $m=0$ (photons),

or (ii) $p \ll mc$ for all occupied levels. (This holds for all particles other than photons at the temperatures that actually have to be considered.)

In the first case we have

$$\alpha = cp \quad \text{(photons)}. \tag{7·21}$$

In the second case, in excellent approximation,

$$\alpha = mc^2 + \frac{p^2}{2m}, \tag{7·22}$$

and the rest energy mc^2 can be dropped, because it is constant and the zero point of energy is irrelevant.

The number (7·19) could have been obtained, by an entirely equivalent procedure, from wave-mechanics. From an asymptotic formula due to H. Weyl the number of proper vibrations with wave-length $>\lambda$ is for any wave-motion confined to a volume V by any linear boundary conditions,

$$\frac{4\pi}{3}\frac{V}{\lambda^3}, \tag{7·23}$$

multiplied by a small integer 1, 2, or 3, depending on the tensor character of the waves, which determines the different polarizations a plane wave can exhibit. 'Asymptotic' means that the expression becomes exact in the limit $V/\lambda^3 \to \infty$. With the universal De Broglie relation between momentum and wave-length

$$p = \frac{h}{\lambda}$$

we get from (7·23)

$$\frac{4\pi}{3}\frac{V}{h^3}p^3,$$

and hence (7·19) for the number between p and $p+dp$.

The two equivalent ways of looking at (7·19), viz. either as counting the number of quantum states of a particle, or as counting the number of wave-mechanical proper vibrations of the enclosure, interest us for this reason. The second attitude makes us think of the 'n_s particles present in state α_s' as of a proper vibration (or a 'hohlraum' oscillator to use a customary expression) in its n_sth quantum level. (This attitude really corresponds to so-called second quantization or field quantization.) n_s becomes a quantum number and the stipulation that the system of quantum numbers

$$n_1, \quad n_2, \quad n_3, \quad \ldots, \quad n_s, \quad \ldots,$$

determines only one state of the gas, not a class of

$$\frac{n!}{n_1!\, n_2! \dots n_s! \dots}$$

states, ceases to be a strange new adoption, and comes into line with the ordinary view about quantum states and their statistical weight (viz. equal for any two of them).

It is the first, the particle attitude, that has suggested the term 'new statistics' which is frequently used. And that is why this idea of new statistics did not, originally, arise in connexion with heat radiation, because here the wave point of view was the historical one, the classical one—nobody thought of any other at the outset. The wave picture was considered to be (and historically was) the classical description. The quantization of the waves therefore duly appeared to be a 'first' quantization and nobody thought of anything like 'second quantization'.

Not until the idea of photons had gained considerable ground did Bose (about 1924) point out that we could, alternatively to the 'hohlraum' oscillator statistics, speak of photon statistics, but then we had to make it 'Bose statistics'. Very soon after, Einstein applied the same to the particles of an ideal gas. And thereupon I pointed out that we could also in this case speak of ordinary statistics, applied to the wave-mechanical proper vibrations which correspond to the motion of the particles of the gas.

The wave point of view in both cases, or at least in all *Bose* cases, raises another interesting question. Since in the Bose case we seem to be faced, mathematically, with a simple oscillator of the Planck type, of which the n_s is the quantum number, we may ask whether we ought not to adopt for n_s half-odd integers

$$\tfrac{1}{2},\quad \tfrac{3}{2},\quad \tfrac{5}{2},\quad \dots,\quad n+\tfrac{1}{2},\quad \dots,$$

rather than integers. One must, I think, call that an open dilemma. From the point of view of analogy one would very much prefer to do so. For, the 'zero-point energy' $\tfrac{1}{2}h\nu$ of a Planck oscillator is not only borne out by direct observation in the case of crystal lattices, it is also so intimately linked up with the Heisenberg uncertainty relation that one hates to dispense with it. On the other hand, if we adopt it straightaway, we get into serious trouble, especially on contemplating changes of the volume (e.g. adiabatic compression of a given volume of black-body radiation), because in this process the (infinite) zero-point energy seems to change by infinite amounts! So we do not adopt it, and we continue to take for the n_s the integers, beginning with 0.

After this digression let us go back to our problem. We shall not treat the photon case for the moment; it is too well known and the reader will easily supplement it for himself. So we use (together with (7·19)) (7·22), where we drop the irrelevant constant mc^2. That gives

$$\frac{4\pi V}{h^3}p^2dp, \text{ single-particle states } p, p+dp,$$

$$\alpha = \frac{p^2}{2m} = \text{kinetic energy of single particle.}$$

Using this we transcribe the sums in (7·14), (7·15) and (7·16) into integrals; whereby we immediately introduce throughout the dimensionless integration variable

$$x = p\sqrt{\frac{\mu}{2m}} = p/\sqrt{(2mkT)},$$

so that the integrals are reduced to functions of the one parameter ζ; we obtain

$$n = \frac{4\pi V}{h^3}(2mkT)^{\frac{3}{2}}\int_0^\infty \frac{x^2\,dx}{\frac{1}{\zeta}e^{x^2}\mp 1}, \tag{7·24}$$

$$\Psi = k\log Z = -nk\log\zeta \mp \frac{4\pi Vk}{h^3}(2mkT)^{\frac{3}{2}}\int_0^\infty \log\left(1\mp\zeta e^{-x^2}\right)x^2\,dx, \tag{7·25}$$

$$U = \frac{4\pi V}{h^3}(2m)^{\frac{3}{2}}(kT)^{\frac{5}{2}}\int_0^\infty \frac{x^4\,dx}{\frac{1}{\zeta}e^{x^2}\mp 1}. \tag{7·26}$$

As is seen by a glance, the first of these equations (which determines ζ as a function of $(V/n)\,T^{3/2}$) expresses the fact that the partial derivative of Ψ with respect to ζ vanishes.

Notice also that by partial integration of the integral containing the logarithm the following alternative expression for Ψ can be obtained:

$$\Psi = k\log Z = -nk\log\zeta + \frac{8\pi Vk}{3h^3}(2mkT)^{3/2}\int_0^\infty \frac{x^4\,dx}{\frac{1}{\zeta}e^{x^2}\mp 1}. \tag{7·27}$$

From this (using the remark about $\partial\psi/\partial\zeta = 0$) you easily confirm (7·26) by forming

$$U = T^2\frac{\partial\Psi}{\partial T}, \tag{7·28}$$

and you could just as easily calculate the pressure

$$p = T\frac{\partial\Psi}{\partial V}. \tag{7·29}$$

But you see at once from (7·24) and (7·25), first that ζ, and then, therefore, that Ψ is only a function of $VT^{3/2}$ (with n constant). From this and the two preceding equations you easily infer

$$pV = \tfrac{2}{3}U \tag{7·30}$$

in both cases, and, by the way, also in the classical theory of an ideal monatomic gas (for heat radiation $pV = \frac{1}{3}U$; that means that p is comparatively much greater, because there U is the total energy, while here it is only the kinetic energy). Another general relation can now be read off (7·27), viz.

$$\Psi = -nk\log\zeta + \frac{pV}{T},$$

and since it equals

$$S - \frac{U}{T},$$

$$nkT\log\zeta = U - TS + pV; \tag{7·31}$$

that is to say, $nkT\log\zeta$ is the thermodynamic potential. (Again an instance of a mathematical auxiliary quantity acquiring physical meaning! At the same time this confirms the fact that our present considerations are not simply an application of the physical method of Darwin and Fowler. For in the latter method $\log z$ was $-1/kT$.)

CHAPTER VIII

EVALUATION OF THE FORMULAE. LIMITING CASES

To determine the actual behaviour of such a degenerate gas requires the numerical evaluation of the two definite integrals for varying ζ. We indicate the general plan of this work.

First, from (7·24), viz.

$$1 = \frac{4\pi(2mk)^{\frac{3}{2}}}{h^3} \frac{V}{n} T^{\frac{3}{2}} \int_0^\infty \frac{x^2\,dx}{\frac{1}{\zeta}e^{x^2} \mp 1}, \tag{8·1}$$

we get the functional relation between

$$\frac{V}{n} T^{\frac{3}{2}} \quad \text{and} \quad \zeta. \tag{8·2}$$

Then, from (7·26) (and (7·30)) we get

$$\frac{2}{3}\frac{U}{nkT} = \frac{pV}{nkT} = \frac{2}{3}\frac{4\pi(2mk)^{\frac{3}{2}}}{h^3} \frac{V}{n} T^{\frac{3}{2}} \int_0^\infty \frac{x^4\,dx}{\frac{1}{\zeta}e^{x^2} \mp 1}. \tag{8·3}$$

The latter gives us the departure from the ordinary gas laws, for it is just 1 for them. Indeed, if we divide (8·3) by (8·1), member by member,

$$\frac{2}{3}\frac{U}{nkT} = \frac{pV}{nkT} = \frac{2}{3}\frac{\displaystyle\int_0^\infty \frac{x^4\,dx}{\frac{1}{\zeta}e^{x^2} \mp 1}}{\displaystyle\int_0^\infty \frac{x^2\,dx}{\frac{1}{\zeta}e^{x^2} \mp 1}}. \tag{8·4}$$

Now for ζ very small we get for the ratio of the two integrals

$$\frac{\displaystyle\int_0^\infty e^{-x^2} x^4\,dx}{\displaystyle\int_0^\infty e^{-x^2} x^2\,dx} = \frac{3}{2}. \tag{8·5}$$

So ζ very small gives (in both cases) the classical behaviour. (ζ is also called the parameter of degeneration.) Both integrals are then very small, and that means, from (8·1),

$$\frac{V}{n} T^{\frac{3}{2}} \quad \text{very large.}$$

That is: high temperature, low density. This is at once (*a*) satisfactory, (*b*) disappointing, (*c*) astonishing.

(*a*) It is satisfactory because we have to find the classical behaviour for high temperature and low density (at least in the Bose case) in order not to contradict old, well-established experimental evidence.

(*b*) The densities are so high and the temperatures so low—those required to exhibit a noticeable departure—that the van der Waals corrections are bound to coalesce with the possible effects of degeneration, and there is little prospect of ever being able to separate the two kinds of effect.

(*c*) The astonishing thing is that the 'new statistics' which replaces just by 1 the factor

$$\frac{n!}{n_1!\,n_2!\ldots n_s!\ldots}$$

(very large in 'the old one', indeed its outstanding feature) should ever give the same behaviour as the old one (if at all, one might expect this rather at $T\to 0$, where the factor *would* approach to 1 in the old theory!).

The solution of this paradox is, that this factor when worked out, applying classical statistics to the quantum levels of the single particles, is not just 1 but $n!$. And that 'does no harm', because it is constant (the harm it does work after all we shall see presently). In other words, the quantum cells are, at high temperatures and with low density, so numerous, that on the average, even in the 'most populated region', only every 10,000th or 100,000th is occupied at all. The n_s are either zero (most of them) or 1, hardly ever 2. And that is why it makes no difference whether the latter possibility is either excluded (Fermi-Dirac) or endowed with a greater statistical weight (Bose-Einstein)—it is negligible anyhow.

The above contention about the occupation numbers is made good by the following considerations. We recall the expression for the average of the occupation number n_s (7·16)

$$\bar{n}_s = \frac{1}{\frac{1}{\zeta}e^{\mu\alpha_s}\mp 1}. \qquad (8\cdot 6)$$

Now for $\zeta \ll 1$, since $e^{\mu\alpha_s} > 1$, we can omit the ∓ 1 and have

$$\bar{n}_s = \zeta e^{-\mu\alpha_s} \leqslant \zeta,$$

showing at once that $\bar{n}_s \ll 1$ when ζ is, and that proves the contention. Moreover, since in the 'really most populated' region, which is $\mu\alpha_s = \alpha_s/kT \approx 1$, the exponential is still of the order of unity (not smaller), we can say that

$$\bar{n}_s \approx \zeta \tag{8·7}$$

gives the true order of magnitude also in the truly interesting region. It is worth while to inquire just how small it is! (I maintained above that it was about 1/10,000th or 1/100,000th.)

That we easily get from (8·1)

$$1 = \frac{4\pi(2mk)^{3/2}}{h^3}\frac{V}{n}T^{3/2}\,.\,\overbrace{\zeta\frac{\sqrt{\pi}}{4}}^{\text{value of the integral}},$$

or

$$\frac{1}{\zeta} = \left(\frac{2\pi mkT}{h^2}\right)^{3/2}\frac{V}{n}. \tag{8·8}$$

This is expected to be a large number. Let us compute it for normal conditions (0° C. and 1 atm.) and for helium, the lightest monatomic gas, taking for convenience 1 mol.:

$$\begin{array}{ll}
\log 2\pi & = 0{\cdot}79818 \\
\log m_{\mathrm{H}} & = 0{\cdot}22337-24 \\
\log 4 & = 0{\cdot}60206 \\
\log k & = 0{\cdot}14003-16 \\
\log 273{\cdot}16 & = 2{\cdot}43642 \\
\hline
 & \ 0{\cdot}20006-36 \\
 & \ 0{\cdot}64226-53 \\
\hline
 & \ 0{\cdot}55780+16 \\
 & \ 0{\cdot}27890+\ 8 \\
\hline
 & \ 0{\cdot}83670+24
\end{array}
\qquad
\begin{array}{ll}
\log h & = 0{\cdot}82113-27 \\
\log h^2 & = 0{\cdot}64226-53 \\
\hline
 & \\
\log V & = 4{\cdot}35054 \\
\log n & = 23{\cdot}77973 \\
\hline
 & \ 0{\cdot}57081-20 \\
 & \ 0{\cdot}83670+24 \\
\hline
 & \ 5{\cdot}40751 \\
\text{Number} & \ 255{,}570.
\end{array}$$

N.B.

$$\left.\begin{array}{ll}
m_{\mathrm{H}} & = 1{\cdot}6725\times 10^{-24} \\
k & = 1{\cdot}3805\times 10^{-16} \\
h & = 6{\cdot}6242\times 10^{-27} \\
V & = 2{\cdot}2415\times 10^{4} \\
n & = 6{\cdot}0228\times 10^{23}
\end{array}\right\}\ \text{c.g.s.° C.}$$

Hence, in these conditions

$$\frac{1}{\zeta} = 255{,}570 \quad \text{(pure number)}. \tag{8·9}$$

The occupation would remain extremely scarce even under strong compression and considerably lower temperature (see (8·7) and (8·8)).* But at the same time we can estimate that if a compression to about 1/100th the volume and a refrigeration to about 1/100th the temperature (thus to 2–3° K.) could be performed without liquefaction, that would give a factor 1/100,000th, and we would just reach the region where ζ ceases to be 'very small'. So the region of noticeable gas degeneration is by no means outside the reach of experiment, only (as I said) its effects are inextricably mixed up with the 'van der Waals corrections'.

The entropy constant. But eqn. (8·8) has also a direct and important application to experiment, viz. for computing the so-called entropy constant or chemical constant, or, to put it more concretely, the vapour-pressure formula of an ideal gas. And that it gives it correctly (while the classical theory gives pure nonsense) is the true justification of the new point of view.

Remember that we had found

$$nkT\log\zeta = \text{thermodynamic potential} = U - TS + pV, \tag{8·10}$$

from which the entropy

$$S = nk\log\left(\frac{1}{\zeta}\right) + \frac{U+pV}{T} = nk\log\frac{1}{\zeta} + \tfrac{5}{2}nk. \tag{8·11}$$

(Thus in (8·9) we have virtually computed the entropy; that is why I took the trouble to compute it exactly instead of merely estimating it.) But we are now interested in the general connexion, and using (8·8) we get

$$S = nk\log\left(\frac{V}{n}\right) + \tfrac{3}{2}nk\log T + nk\log\left(\frac{2\pi mk}{h^2}\right)^{\frac{3}{2}} + \tfrac{5}{2}nk. \tag{8·12}$$

Please note in the first place that this expression is sound, as

* Notice that the relative fluctuation of these small occupation numbers $n_s \sim \zeta$ is extremely great, viz. $1/\sqrt{\zeta} \sim 500$ or 50,000 %.

regards the dependence on V and n; if you increase n and V proportionally, S takes up the same factor. That may seem trivial, but that is just its first and supreme merit—it is just the point in which the classical point of view pitiably fails, as we shall see.

After having taken due notice of this soundness, we now refer to 1 gram-molecule, so that $nk = \mathbf{R}$, the gas constant. In the argument of the first log we supply a $1/k$ (correcting for it in the constant) and then use

$$\frac{V}{nk} = \frac{V}{\mathbf{R}} = \frac{T}{p},$$

because it is more usual to speak of the pressure than of the volume in this connexion (viz. in the case of the saturated vapour, to which we shall immediately proceed). Then

$$S = -\mathbf{R}\log p + \tfrac{5}{2}\mathbf{R}\log T + \mathbf{R}\log\frac{(2\pi m)^{\frac{3}{2}}k^{\frac{5}{2}}}{h^3} + \tfrac{5}{2}\mathbf{R}.$$

If we now measure $\mathbf{R}$ in cal./°C. and equate this S to the actual heat supply per gram-molecule on evaporation Λ_p (which must be taken from experiment),* divided by T, thus

$$S = \frac{\Lambda_p}{T}, \tag{8·13}$$

we get the famous Sackur-Tetrode vapour-pressure formula, valid for temperatures low enough to allow us to neglect the entropy of the condensed state. To higher temperatures, where this will no longer be so, and the gas will cease to behave 'monatomically' by rotations and oscillations coming in, we proceed by following up either theoretically or experimentally the specific heats of both the gas and the condensed state, which inform us of all further changes of the respective entropies and of the heat of evaporation. All that is then classical thermodynamics, and it is well known how from vapour pressures we

* The index p is a reminder that 'p is constant' on evaporation. The heat supply *includes* that part which makes up for the work pV $(= \mathbf{R}T)$ done on evaporation.

can predict all chemical equilibria which include a gaseous phase. The salient point was to find the value of the constant for every gas—depending only on the mass of the particle: for this makes it possible to predict such equilibria from pure caloric or energetic measurements (without having to set up even *one* equilibrium experimentally).

The heat of evaporation at some temperature must, of course, be procured by experiment, but that too is only the measurement of an *energy difference*, which does not necessarily require the actual performance of a reversible transition. One can obtain that energy difference at any temperature and then compute it for any other from calorimetric measurements; one can obtain it by any means, including, for example, explosions within a bomb in a calorimeter; one may even estimate it theoretically from any knowledge one may have of the forces which keep the atoms in the crystal (forming the condensed) together.

But one must beware of one suggestive error. One is tempted to say: 'Well, at those low temperatures which we are speaking of, the energy of the solid is practically zero, the vapour behaves as an ideal (non-degenerate) monatomic gas, so its energy is $\frac{3}{2}\mathrm{R}T$ and the heat of evaporation is thus $\frac{3}{2}\mathrm{R}T+\mathrm{R}T = \frac{5}{2}\mathrm{R}T$.'

This would be a mistake, suggested by current terminology, though it would hardly be fair to put the blame upon it. What happens is this. In saying the energy of the condensed is zero* and in saying the energy of the vapour is $\frac{3}{2}\mathrm{R}T$, we are not using the same zero level of energy. We give no credit to the particles in the gas for having extricated themselves from each other's spheres of attraction. It is this part of the heat of evaporation which, of course, cannot possibly be indicated by any general theory.

The failure of the classical theory. Gibbs's paradox. Let us glance at corresponding classical considerations, which on super-

* This, by the way, would in itself not be quite correct, because there is a considerable zero-point vibration in the crystal.

ficial inspection seem to give almost the same results. We have merely to repeat the considerations of Chapter IV on the free mass-point. There we only aimed at the elementary classical results for ideal gases, we took the phase volume itself as a measure of the number of quantum states and we did not at all evaluate the value of the additive constant in $\log Z$. Supplying now the factor h^{-3} we get for the sum-over-states of the single mass-point:

$$Z_{\text{single}} = \frac{V}{h^3}(2mkT)^{\frac{3}{2}} \underbrace{\iiint\limits_{-\infty}^{+\infty} e^{-(\xi^2+\eta^2+\zeta^2)}\,d\xi\,d\eta\,d\zeta}_{(\sqrt{\pi})^3}$$

$$= \frac{V}{h^3}(2\pi mkT)^{\frac{3}{2}},$$

$$\Psi_{\text{single}} = k\log Z_{\text{single}} = k\log V + \frac{3k}{2}\log T + k\log\left(\frac{2\pi mk}{h^2}\right)^{\frac{3}{2}}. \tag{8·14}$$

Hence for the n mass-points forming the gas (according to the principles laid down there, sound in themselves)

$$\Psi_{\text{gas}} = nk\log V + \frac{3nk}{2}\log T + nk\log\left(\frac{2\pi mk}{h^2}\right)^{\frac{3}{2}}. \tag{8·15}$$

Now, quite in general $\Psi = S - U/T$, and since in our case $U = \frac{3}{2}nkT$, we get

$$S_{\text{gas}} = nk\log V + \frac{3nk}{2}\log T + nk\log\left(\frac{2\pi mk}{h^2}\right)^{\frac{3}{2}} + \tfrac{3}{2}nk. \tag{8·16}$$

Let me first point out an objection, that must not be raised, viz. that this goes to $-\infty$ for $T \to 0$, instead of going to 0. This objection would be as little justified as in the case of (8·12). The (8·16) does not claim to be right at very low temperature, because the mere *counting* of quantum levels is then no longer sufficient. On the contrary, we made sure that the present 'Boltzmannian' point of view must, just as the other, lead to $S = 0$ for $T \to 0$, when all the particles relapse into their lowest state. Indeed

$$\frac{n!}{n_1!\,n_2!\ldots n_s!\ldots} \tag{8·17}$$

then goes to 1, and its logarithm to zero.

The true blemish on (8·16), which renders it absolutely unusable in spite of its great similarity to (8·12), is that the dependence on n and V is unsound. The constant is not doubled when n and V are doubled. But it is doubled, when n alone is doubled. If you try to use it, as before, to determine the vapour-pressure formula, you get the bewildering experience, not perhaps that you get the vapour pressure wrong, but you do not get it at all! Indeed, call λ the heat of evaporation per particle and put (8·16) equal to

$$\frac{n\lambda}{T},$$

then you can cancel n throughout and what is then determined (given the temperature) is not the vapour *pressure*, but the vapour *volume*, the absolute volume of the vapour, independent of the number n of particles it contains. Given this 'correct' volume any amount of liquid could evaporate into it, or vice versa, without disturbing the equilibrium!

Comparing (8·12) and (8·16) closely we find that the superabundant additive term in the latter reads

$$nk\log n - nk = kn(\log n - 1) = k\log n!. \qquad (8\cdot18)$$

That is clearly k times the logarithm of the value the 'permutation' (8·17) takes when all the n_s are either zero or 1. That shows that the 'new statistics' avoid or amend the fault of the old indeed by their essential step, viz. 'not counting permutations' and therefore taking that factor always = 1. (It was mentioned above, p. 54, that this $n!$ did no harm in the equation of state, etc., but that the harm it did after all would presently appear.)

But there are a few more quite interesting connexions. The superabundant addendus (8·18) has the consequence that if you join, say, two gramme-molecules of gas together, without doing anything else, the entropy is not doubled, but there is (as may easily be verified) an additional increase of

$$2R\log 2.$$

1 mol.	1 mol.

Now, this is very interesting. For that is exactly the increase of entropy which arises when you join two gramme-molecules of two different, but not chemically reacting, gases in the same way, take out the diaphragm separating them—and wait. For then diffusion sets in and we know that eventually the increase of entropy for each of the gases is the same as if it had been allowed to occupy double the volume by itself.

It was a famous paradox pointed out for the first time by W. Gibbs, that the same increase of entropy must not be taken into account, when the two molecules are of the same gas, although (according to naïve gas-theoretical views) diffusion takes place then too, but unnoticeably to us, because all the particles are alike. The modern view solves this paradox by declaring that in the second case there is no real diffusion, because exchange between like particles is not a real event—if it were, we should have to take account of it statistically. It has always been believed that Gibbs's paradox embodied profound thought. That it was intimately linked up with something so important and entirely new could hardly be foreseen.

After a railway accident, or a fire, or a similar disaster, the authorities are always anxious to answer the question: How could it have happened?

How could it have happened that even with a wrong gas model we arrived at the non-additive result (8·16) for the entropy? It will be remembered that in our quite general development of the theory we took the most anxious precautions that the log of the sum-over-states and thereby all thermodynamic functions were strictly additive. Even this very formula with which the disaster happened was deduced from log Z_{single} by just multiplying it by n. How could it then not be proportional to n?

Well, look at it—it is, of course, proportional to n, but with the volume constant. What happens now if, say, we double the volume as well? *We inadvertently change the allowed quantum states of the single particle, for we double their density all along the energy line.* But in our general considerations, by which we proved that when two systems A and B (with quantum levels α_m and β_k) are joined together, their thermodynamic functions are additive, we had tacitly assumed that the α_m and β_k are not changed by joining the two and that therefore the combined system has the quantum levels

$$\epsilon_l = \alpha_m + \beta_k.$$

That explains the failure. It is true that also in the new theory this peculiar change of the single-particle levels (on putting two bodies of the same gas together) subsists. But here the single levels play only the role of an auxiliary conception, a convenient means of indicating the gas levels. As regards the latter, the requirement that the levels of the compound system should be those resulting additively from the levels of the constituents in all combinations is not rigorously fulfilled,* but obviously in sufficient approximation to make all thermodynamic functions additive.

Digression: Annihilation of matter? This is perhaps the best moment to speak of an interesting aspect of the relation (8·1) which arises when we drop the condition $\sum_s n_s = \text{const.}$ in the case of particles with non-vanishing rest-mass, allowing, as it were, particles to be created or annihilated in collisions, the balance $\pm mc^2$ going to or from the account of the kinetic energy. One must in this case, of course, use the non-truncated expression (7·22)

$$\alpha = mc^2 + \frac{p^2}{2m}, \tag{7·22}$$

and it is easily seen that this just yields an additional factor $e^{mc^2/kT}$ to $1/\zeta$, both in (8·1) and in (8·2). If we did not drop the condition for $\sum_s n_s$ the effect would be nil—an irrelevant change

* On joining up, the single-particle level schemes are superposed, as it were. All combinations of a set $n_1, n_2, \ldots, n_s, \ldots$ with a set $n_1', n_2', \ldots, n_s', \ldots$ are, of course, levels of the combined system. But there are others in addition, because after joining up, the sums $\sum_s n_s$ and $\sum_s n_s'$ are no longer required to be separately constant, but only their sum.

in the definition of ζ (that is why we could delete mc^2 in the preceding). But now we do drop it. Then $\zeta = 1$. (8·1) determines no longer ζ, but n, or n/V, the particle density. Moreover, formally (8·1) is the same as before, but with $e^{-mc^2/kT}$ in place of ζ. For all temperatures in question this is an extremely small number, much smaller than the one indicated in (8·9), viz. $\zeta = 1/255{,}570$. We must therefore be prepared to find that at ordinary temperature the dropping of the condition would entail extreme rarefication. We get, in complete analogy with (8·8),

$$e^{mc^2/kT} = \left(\frac{2\pi mkT}{h^2}\right)^{\frac{3}{2}} \frac{V}{n}. \qquad (8{\cdot}19)$$

(If we take logarithms $\times nk$, we get formally, in full analogy with (8·12) and (8·13), the vapour-pressure formula of a substance with the enormous heat of evaporation nmc^2, divided by the mechanical equivalent of the calorie.) As an example, let us use the case of helium under normal conditions, of which we have procured accurate data before. The number per cm.³, n/V, will obviously now be smaller* by the factor

$$255{,}570 e^{-mc^2/273k}. \qquad (8{\cdot}20)$$

I have worked out the exponential and find

$$10^{-6{\cdot}9343\ldots \times 10^9}. \qquad (8{\cdot}21)$$

The other factor 255,570 must be neglected, since the exponent in (8·21) is not computed to 9 decimal places. Within the accuracy reached, it is even irrelevant whether we speak of the density in gram-molecules per litre or in single particles per universe, because that means only a factor of about 10^{110}.

The result is typical of what one gets when any other possibilities for the annihilation of matter, e.g. transition into heat radiation are taken into account as well. Unless we want to assume that these kinds of transition are impossible, we are astonished that there is so much ponderable matter left in the universe as there is. The only way out seems to be to assume that the transition is a very slow process and that not very far

* Smaller than 6×10^{23} per 22 litres.

back the conditions of the universe were very different from what they are now.

Digression on the uncertainty relation. Before we consider the case of degeneration proper (i.e. $\zeta \not\ll 1$) let us consider from another point of view the quantity on the right of (8·8)

$$\frac{1}{\zeta} = \left(\frac{2\pi mkT}{h^2}\right)^{\frac{3}{2}} \frac{V}{n}, \tag{8·8}$$

which, it will be remembered, is the form (8·1) takes for $\zeta \ll 1$. A noticeable departure from classical behaviour will begin at temperatures and volumes where the second member is no longer a very large number, but becomes comparable with unity. Still nearer to unity is, then, its third root which is

$$\frac{\sqrt{(2\pi mkT)}}{h} \sqrt[3]{\frac{V}{n}}; \tag{8·22}$$

it allows of a very simple interpretation. For since the average value of the energy is certainly of the order

$$\frac{\overline{p^2}}{2m} \sim \tfrac{3}{2}kT,$$

the average momentum square is of the order

$$\overline{p^2} \sim 3mkT.$$

The square root of this is certainly an upper limit to the uncertainty of momentum, or rather, in a way, it is precisely the uncertainty about the momentum of a particle picked out at random.* Hence from Heisenberg's uncertainty relation a lower limit of the uncertainty Δx in the location of the particle is

$$\Delta x > \frac{h}{2\pi \sqrt{(3mkT)}}.$$

Thus (8·22) is, as to order of magnitude, the ratio between the

* Whether this be granted or not, the statement about the upper limit is undeniable. For if the uncertainty were greater, $\overline{p^2}$ would be greater, and that would mean that the temperature is higher than it actually is.

average distance between the particles $\sqrt[3]{(V/n)}$, and the maximum accuracy with which a particle can be located,

$$\frac{1}{\Delta x}\sqrt[3]{\frac{V}{n}}. \tag{8·23}$$

When this is no longer large, when it becomes of the order of 1, one has, I think, to say that the particles become entirely blurred, the particle aspect breaks down, and one is no longer allowed to speak of a granulated structure of matter.

This remark is of much wider application, far beyond our present subject-matter, and I dare say it is directly supported by experiment. Wherever the particle aspect truly enters the interpretation of an experiment (e.g. in Wilson chamber experiments or in the counting of cosmic-ray particles) it is with corpuscles in extreme rarefication and of high speed. For, the actual momentum always sets an upper limit to the uncertainty of momentum, and thus sets a limit to the accuracy of location, and to the crowding of similar particles, if the crowding is not to prevent altogether their being recognizable as separate particles.

But what about the crowding of particles in liquids and solids? The volume per particle V/n is here roughly about 1000 times smaller than in a gas under standard conditions. Hence, if we contemplate a crystal at a temperature about 100 times smaller (thus between 2 and 3° K.), these two circumstances seem to lower the value of (8·8) by a factor of roughly one-millionth, as compared with the numerical value (8·9). There is a certain compensation by the *mass* m, if it is larger than $4m_{\mathrm{H}}$ (remember that (8·9) was computed for helium gas). Yet the situation seems to be, on the whole, unfavourable to the particle aspect. Is then the point of view put forward in the preceding paragraphs not flatly contradicted by the fact, that the particle models of the structure of crystal lattices most certainly do not break down at very low temperature—in fact, quite the contrary?

There is, I think, no contradiction, for two reasons: First, it follows both from theory and from experimental evidence* that the vibrational energy of the crystal particles, while it becomes more and more independent of temperature, yet does not approach to zero, as T goes towards zero. It approaches, as regards order of magnitude, to $k\theta$ per particle, where θ is the so-called Debye temperature, a parameter which is used in the theoretical description of the rapid falling of the specific heat from its Dulong-Petit value to practically zero, and which indicates roughly the region of the main slope of the curve. This θ is always much larger than the low value of T we envisaged above. It ranges from 88 (Pb) to about 2000 (C), and there is an intrinsically understandable tendency for it to be small only for heavy atoms or particularly wide spacing in the crystal (e.g. K, with $\theta = 99$, spacing 4·5 Å.), but high for light atoms. (This makes for a further compensation in our expression (8·8), in which T is now to be replaced by θ and which, of course, is only to indicate the order of magnitude.)

But there is a second point that ought to be mentioned. In spite of the usefulness of the lattice models, the great successes in the thermodynamics of crystals involve the wave aspect, not the particle aspect. They were reached by P. Debye, in his theory of the specific heat of solids, mentioned above, by attributing certain quantum levels not to the single particles, but to the proper vibrations of the lattice as a whole. (This seemed, at the time, to be a most perplexing step!) At very low temperature the energy content and the specific heat of a crystal are expressed by formulae which are, in their derivation, almost identical replicas of those which hold for black-body radiation. I am referring to the famous T^3 or T^4 law.

The very accurate location of atoms within a crystal by X-ray methods (I mean the measuring of lattice spacing and of so-

* Concerning the intensity distribution in Laue photographs, taken at very low temperature; the spots in the rear, scattered at obtuse angles, remain *weaker* than those in front, scattered under small angles with the incident ray.

called 'parameters'), if looked upon as locating individual atoms, far surpasses the limits of accuracy drawn by the uncertainty relation. But we must not regard those data as referring to any single atom. The very accurate measurement of those distances is only made possible by, and depends entirely on, the fact that they repeat themselves millions and millions of times throughout the crystal, in much the same way as the distance between successive wave crests is repeated again and again throughout a wave. Indeed, I deem the whole lattice structure to be something very akin to a standing De Broglie wave. It could, I think, be treated as such, but the task is extremely intricate, on account of the very strong interaction between these waves. (The current attitude is to treat the interactions as forces between particles, to build up the crystal lattice along the lines of the particle aspect and then to contemplate—and to quantize—the sound waves, set up in this lattice, which have only a very weak interaction.) Yet certain connexions are recognizable even now. For example, the secondary beams in the X-ray-diffraction pattern are determined by the quanta of momentum which can be imparted to the light wave by the crystal lattice, by virtue of its periodic structure, when it is regarded as a standing wave. (This is not a new mathematical theory of X-ray diffraction; it is an alternative interpretation which the current theory admits.)

Gas-degeneration proper. The quantitative study of the deviations from the classical gas laws which occur when ζ is not very small are not of great practical interest, except for one case, the theory of the electrons in metals. But we ought to indicate briefly the mathematical methods of coping with the task; these are very simple and easy of application to all cases that are of any interest.

First, there is the case of *weak* degeneration—ζ small but not very small—the first deviations from the classical laws to be expected on increasing the density and lowering the temperature. Even though, as I have said, they are bound to be mixed up with other influences, it is of interest to know what part of the observed deviations could be attributed to the new ideal gas laws. Secondly, there is the case of *strong* degeneration which in the Fermi case covers precisely the electron theory of metals (Sommerfeld, *Z. Phys.* vol. 47, 1928). In the Bose case it is linked with the 'phenomenon of Bose-Einstein condensation' which is, at any rate, of great theoretical interest—a most unexpected discontinuous behaviour of the sum-over-states, and therefore of the material system, which we shall discuss in detail.

Weak degeneration. If we envisage the equations (8·1) and (8·4) which contain the laws in the concisest form

$$1 = \frac{4\pi(2mk)^{\frac{3}{2}}}{h^3}\frac{V}{n}T^{\frac{3}{2}}\int_0^\infty \frac{x^2\,dx}{\frac{1}{\zeta}e^{x^2}\mp 1}, \tag{8·1}$$

$$\frac{2}{3}\frac{U}{nkT} = \frac{pV}{nkT} = \frac{2}{3}\frac{\displaystyle\int_0^\infty \frac{x^4\,dx}{\frac{1}{\zeta}e^{x^2}\mp 1}}{\displaystyle\int_0^\infty \frac{x^2\,dx}{\frac{1}{\zeta}e^{x^2}\mp 1}}, \tag{8·4}$$

we see that we need the two integrals as functions of ζ. Now for ζ not too large (in point of fact for $\zeta \leqslant 1$), one can use the development

$$\frac{1}{\frac{1}{\zeta}e^{x^2}\mp 1} = \frac{\zeta e^{-x^2}}{1\mp\zeta e^{-x^2}} = \zeta e^{-x^2}(1\pm\zeta e^{-x^2}+\zeta^2 e^{-2x^2}\pm\ldots), \tag{8·24}$$

and then integrate term by term. The result reads, if one abbreviates the integrals as I_2 and I_4 respectively,

$$\left.\begin{aligned} I_2 &= \frac{\sqrt{n}}{4}\left(\zeta\pm\frac{\zeta^2}{2^{\frac{3}{2}}}+\frac{\zeta^3}{3^{\frac{3}{2}}}\pm\frac{\zeta^4}{4^{\frac{3}{2}}}+\ldots\right), \\ I_4 &= \frac{3}{2}\frac{\sqrt{n}}{4}\left(\zeta\pm\frac{\zeta^2}{2^{\frac{5}{2}}}+\frac{\zeta^3}{3^{\frac{5}{2}}}\pm\frac{\zeta^4}{4^{\frac{5}{2}}}+\ldots\right). \end{aligned}\right\} \tag{8·25}$$

To get information about the beginning degeneration, one would have to insert the first series in (8·1), reverse it step by step in a well-known manner, and thus get ζ developed by powers of the small number

$$\frac{h^3}{4\pi(2mk)^{\frac{3}{2}}}\frac{n}{V}T^{-\frac{3}{2}}.$$

This development one would insert in (8·4), after having replaced its second member by the power series

$$\frac{2}{3}\frac{I_4}{I_2} = 1 \mp \frac{1}{2^{\frac{5}{2}}}\zeta + \dots, \tag{8·26}$$

obtained from (8·25) by dividing the second power series by the first. We are not interested in following this up in any detail. Convergence becomes poorer as ζ increases but actually subsists till $\zeta = 1$. (For $\zeta = 1$, by the way, the series represent, apart from simple factors, the Riemannian ζ function of argument $\frac{3}{2}$ and $\frac{5}{2}$ respectively, which can be found in tables.)*

So much for weak degeneration.

Medium degeneration. This case has not come to be of any practical interest. Mathematically it is characterized, of course, by poor convergence of both the series we have derived for weak, and the one we shall derive for strong degeneration. I shall use this only as an occasion to point out a slight simplification, which holds for any $\zeta \leqslant 1$.

We might think that we have to evaluate numerically as functions of ζ four integrals, viz. the one with x^2 and the one with x^4 in both the Bose and the Fermi case. But actually they can be reduced to two. Not that the one with x^4 is reducible to the one with x^2, but the Fermi and Bose functions are reducible to each other. For

$$\frac{1}{\frac{1}{\zeta}e^{x^2}-1} - \frac{1}{\frac{1}{\zeta}e^{x^2}+1} = \frac{2}{\frac{1}{\zeta}e^{2x^2}-1};$$

that gives

$$\frac{1}{\frac{1}{\zeta}e^{x^2}+1} = \frac{1}{\frac{1}{\zeta}e^{x^2}-1} - \frac{2}{\frac{1}{\zeta}e^{2x^2}-1}$$

* See, e.g., Jahnke-Emde, *Tables of Functions*, B.G. Teubner, 1938.

and by iteration

$$\frac{1}{\frac{1}{\zeta}e^{x^2}-1}=\frac{1}{\frac{1}{\zeta}e^{x^2}+1}+\frac{2}{\frac{1}{\zeta^2}e^{2x^2}+1}+\frac{4}{\frac{1}{\zeta^2}e^{4x^2}+1}+ad\,inf.$$

The relations between the integrals can be easily formed from this, when $\zeta \leqslant 1$. (For $\zeta > 1$ the Bose integral becomes meaningless; see below.)

Strong degeneration. Here we must separate the two cases entirely, for extreme degeneration means something entirely different in the Bose-Einstein and in the Fermi-Dirac case. Indeed, since by (8·1) the integral means a count of the particles (see eqn. (8·6)), the integrand must never be negative. Hence with the upper sign (Bose-Einstein) we must have $\zeta \leqslant 1$, and extreme Einstein degeneration is thus characterized by $\zeta = 1$. We shall deal with it in the second place. With the lower sign (Fermi-Dirac) ζ is allowed to exceed 1. Extreme Fermi degeneration is characterized by $\zeta \to \infty$.

(*a*) *Strong Fermi-Dirac degeneration* (lower sign everywhere). The first approximation for ζ very large is easy to obtain, since the characteristic factor of the integrand, viz. the fraction (8·24)

$$\bar{n}_s=\frac{1}{\frac{1}{\zeta}e^{x^2}+1}, \tag{8·24}$$

which, it will be remembered, is the average occupation number of a level α_s $\left(\text{with } x^2=\frac{\alpha_s}{kT}=\frac{p_s^2}{2mkT}\right)$, drops almost abruptly from 1 to 0 at or around that value of x where the fraction is $\frac{1}{2}$, i.e. where

$$x=\sqrt{\log \zeta}. \tag{8·27}$$

Our two integrals take therefore the values

$$I_2=\tfrac{1}{3}(\log\zeta)^{\frac{3}{2}},\quad I_4=\tfrac{1}{5}(\log\zeta)^{\frac{5}{2}}, \tag{8·28}$$

and (8·1) and (8·4) give

$$\left.\begin{aligned} 1 &= \frac{4\pi(2mk)^{\frac{3}{2}}}{h^3}\frac{V}{n}T^{\frac{3}{2}}\tfrac{1}{3}(\log\zeta)^{\frac{3}{2}}, \\ \frac{2}{3}\frac{U}{nkT} &= \frac{pV}{nkT} = \tfrac{2}{5}\log\zeta. \end{aligned}\right\} \tag{8·29}$$

From the first

$$\log\zeta = \left(\frac{3}{4\pi}\right)^{\frac{2}{3}}\frac{h^2}{2mkT}\left(\frac{n}{V}\right)^{\frac{2}{3}}. \tag{8·30}$$

Hence from the second (8·29)

$$p = \frac{2}{3}\frac{U}{V} = \frac{1}{5}\left(\frac{3}{4\pi}\right)^{\frac{2}{3}}\frac{h^2}{m}\left(\frac{n}{V}\right)^{\frac{5}{3}}. \tag{8·31}$$

The last equation contains the complete description of the thermodynamic behaviour of a Fermi gas in the state of extreme degeneration. The most remarkable feature—a necessary consequence of the Nernst theorem—is that the temperature has disappeared from the formula. The gas behaves as a 'pure mechanism'—as indeed every system must, according to the Nernst theorem,* in the limit $T \to 0$. Observe, by the way, that the equation of state, viz.

$$p\left(\frac{V}{n}\right)^{\frac{5}{3}} = \text{const.},$$

is exactly the same as the 'adiabatic relation' for an ideal monatomic gas, at any temperature, in the classical theory.†

That the energy density does not depend on temperature and therefore that the specific heat vanishes, is the basic virtue of this theory in the explanation of the behaviour of the electrons

* Because there are no 'entropy transactions'.

† The 'adiabatic relation' between p and V is the same at all temperatures (and for both the Bose-Einstein and the Fermi-Dirac gas). Indeed, it results from $dQ = dU + p\,dV = 0$ with $pV = \frac{2}{3}U$. But only for the extremely degenerate Fermi-Dirac gas does it coincide with the equation of state.

in a metal. For many years the following points had presented a problem:

(i) The high electric and thermic conductivity of metals point to an electron density of the order of 1 free electron per atom.

(ii) Yet the specific heats of metals obey the Dulong-Petit law at room temperature without any trace of an electronic contribution (which ought to have increased the value by 50 % if the electrons formed a classical ideal gas).

(iii) The electrons emerging from hot metals in the 'Richardson effect' show exactly the Maxwellian distribution of velocities corresponding to that temperature, which seemed to plead strongly for their actually forming a classical ideal gas inside the metal—where, by the way, this same assumption also seemed inevitable for a quantitative explanation of the electric and thermic conductivity, of their famous ratio, and of its temperature coefficient ($= 1/273$).

All these points are satisfactorily explained by the present theory. The expected contribution to the specific heat is removed since U is independent of T. Yet the particles, even at the lowest temperatures, retain considerable velocities, since the Pauli exclusion principle forces them to occupy the n lowest states, of which the highest represent an energy much higher than $\frac{3}{2}kT$. The explanation of the conductivities and of their ratio works out to full satisfaction (and so does the theory of the host of 'effects' as Hall effect, thermo-electricity, etc.). The paradox of the Richardson effect turns out to be a thermodynamic necessity: the 'electron vapour' which the metal gives off, must, by reason of its much lower density, exhibit the properties of a non-degenerate gas at the same temperature, just as, for example, the saturated vapour over a cold crystal is a classical ideal gas, although the atoms inside the crystal may already have been reduced practically to their zero-point energy. The mechanical reason why the electrons emerge in the Richardson effect with an entirely different velocity distribution—much smaller velocities than they have inside the metal—is, that they have to overcome an exit barrier of potential energy, several volts, just like an atom evaporating from a solid or liquid; it is, of course, this potential barrier which takes the role played by the walls of the vessel for an ordinary gas in keeping the electrons together.

That the electron gas is highly degenerate at room temperature, and even at the high temperatures of the Richardson effect, is due to the co-operation of two circumstances: (i) their comparatively high density n/V, about the same as for the atoms of a solid; (ii) their small mass, about 1/2000th of a hydrogen nucleus. This, according to (8·30), produces the high value of $\log \zeta$, required for Fermi degeneration.

The macroscopic properties of metals for which the electrons are responsible—with the single exception, I believe, of diamagnetism and perhaps of supra-conductivity, which we do not yet understand—are not due to the electrons in the densely packed region where all the successive levels are occupied. For there the Pauli exclusion principle forestalls a transition to a neighbouring level, e.g. in the way that an electric field applied to the metal 'from left to right' would cause the (negative) electrons to favour such levels as have a momentum 'from right to left'. For there is no question of choice—'the bus is full', there are no empty seats. Thus we grasp the outstanding importance of that 'region of transition' where the occupation number $\bar{n}_s$ (8·24) changes very abruptly, as I have said, but, nevertheless, continuously from 1 to zero with increasing

$$x\left(=\frac{p_s}{2mkT}\right).$$

It is the region around the x value indicated in (8·27).

That is why a better approximation than we have used above is needed in this application. Even though I do not wish to go into further details of the Sommerfeld theory here, I ought to explain the mathematical procedure. Let us take the integral I_2 as an example. I_4 and other similar integrals that may occur are tackled in exactly the same way. The leading principle is this. In the integrand of

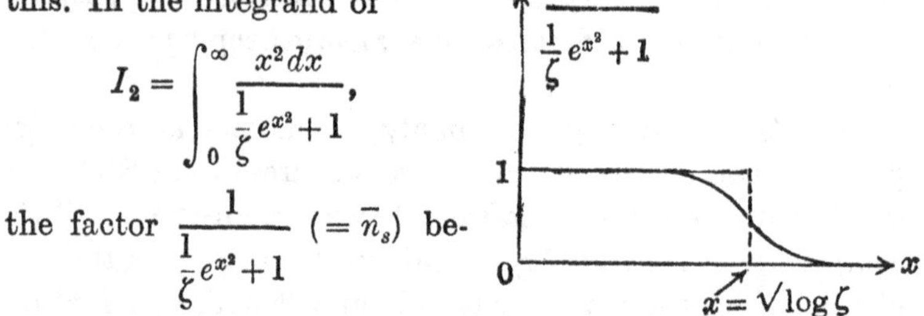

$$I_2 = \int_0^\infty \frac{x^2\,dx}{\frac{1}{\zeta}e^{x^2}+1},$$

the factor $\dfrac{1}{\frac{1}{\zeta}e^{x^2}+1}$ $(=\bar{n}_s)$ behaves as shown in the figure. We have hitherto approximated it by the broken line with ordinates 1 or 0. We continue to take this as the first approximation, but develop the correction in the neighbourhood of the critical abscissa $\sqrt{\log \zeta}$. It is a little more convenient to use the variable

$$u = x^2, \quad \text{with} \quad u_0 = \log \zeta \tag{8·32}$$

the critical abscissa (u is, essentially, the energy); thus

$$\begin{aligned} 2I_2 &= \int_0^\infty \frac{u^{\frac{1}{2}}\,du}{e^{u-u_0}+1} \\ &= \int_{u_0}^\infty \frac{u^{\frac{1}{2}}\,du}{e^{u-u_0}+1} + \int_0^{u_0} \left(\frac{1}{e^{u-u_0}+1} - 1 + 1\right) u^{\frac{1}{2}}\,du \\ &= \int_{u_0}^\infty \frac{u^{\frac{1}{2}}\,du}{e^{u-u_0}+1} - \int_0^{u_0} \frac{u^{\frac{1}{2}}\,du}{e^{u_0-u}+1} + \tfrac{2}{3}u_0^{\frac{3}{2}}. \end{aligned}$$

The last term is the first approximation, the integral taken over the broken line: the other two represent the two 'triangular' surfaces which have to be added and subtracted, respectively, to get the true value. Introduce in both integrals the positive variable t, in the first by $u-u_0 = u_0 t$, in the second by $u_0 - u = u_0 t$. You then get (writing the main term first)

$$2I_2 = \tfrac{2}{3}u_0^{\frac{3}{2}} + u_0^{\frac{3}{2}}\left(\int_0^\infty \frac{dt\,\sqrt{(1+t)}}{e^{u_0 t}+1} - \int_0^1 \frac{dt\,\sqrt{(1-t)}}{e^{u_0 t}+1}\right).$$

We commit a very small error (of relative order $e^{-u_0} = 1/\zeta$), if we also terminate the first integral at $t = 1$. Then we can unite the two. Using the development

$$\sqrt{(1+t)} - \sqrt{(1-t)} = t + \tfrac{1}{8}t^3 + \tfrac{7}{128}t^5 + \ldots,$$

we get

$$2I_2 = \tfrac{2}{3}u_0^{\frac{3}{2}} + u_0^{\frac{3}{2}}\left(\int_0^1 \frac{t\,dt}{e^{u_0 t}+1} + \frac{1}{2}\int_0^1 \frac{t^3\,dt}{e^{u_0 t}+1} + \frac{7}{128}\int_0^1 \frac{t^5\,dt}{e^{u_0 t}+1} + \ldots\right). \tag{8·33}$$

We commit small errors of the same order as before, if we now extend all these integrals to infinity. After that we introduce everywhere the integration variable $u_0 t$, but again call it t. Thus

$$2I_2 = \tfrac{2}{3}u_0^{\frac{3}{2}} + u_0^{-\frac{1}{2}}\int_0^\infty \frac{t\,dt}{e^t+1} + \tfrac{1}{8}u_0^{-\frac{5}{2}}\int_0^\infty \frac{t^3\,dt}{e^t+1} + \tfrac{7}{128}u^{-\frac{9}{2}}\int_0^\infty \frac{t^5\,dt}{e^t+1} + \ldots. \tag{8·33'}$$

Since the integrals are now pure numbers, we have obtained a development in descending powers of the parameter $u_0^2 = (\log \zeta)^2$ which is supposed to be fairly large (at the same time the above neglect of $1/\zeta$ is justified in the circumstances). The integrals are simple numerical multiples of the Riemannian ζ function. For example,

$$\left.\begin{aligned}
&\int_0^\infty \frac{t\,dt}{e^t+1} = \tfrac{1}{2}\zeta(2) = \frac{\pi^2}{12},\\
&\int_0^\infty \frac{t^3\,dt}{e^t+1} = \frac{7\cdot 31}{8}\zeta(4) = \frac{7\pi^4}{120},\\
&\int_0^\infty \frac{t^5\,dt}{e^t+1} = \tfrac{31}{32}5!\,\zeta(6) = \frac{31\pi^6}{252} \quad \text{in general,}\\
&\int_0^\infty \frac{t^p\,dt}{e^t+1} = \left(1-\frac{1}{2^p}\right)p!\,\zeta(p+1),\\
&p \textit{ any} \text{ natural number, not necessarily a prime.}
\end{aligned}\right\} \tag{8·34}$$

The expressions in π come from a formula

$$\zeta(2p) = 2^{2p-1}\frac{\pi^{2p}}{(2p)!}B_p, \tag{8·35}$$

where B_p is the Bernoulli number.

I_4 and similar integrals can be obtained in exactly the same way. We drop the subject here.

(*b*) *Strong Bose-Einstein degeneration.* We have already pointed out that with the lower sign in (8·1) and (8·4) the largest admissible value of ζ is $\zeta = 1$, because the integrand, in virtue of its meaning, must not be negative. In this limiting case, then, we get from (8·1)

$$1 = \frac{4\pi(2mk)^{\frac{3}{2}}}{h^3}\frac{V}{n}T^{\frac{3}{2}}\int_0^\infty \frac{x^2\,dx}{e^{x^2}-1}. \tag{8·1'}$$

The integral is a pure number, moreover (see (8·25)),

$$\int_0^\infty \frac{x^2\,dx}{e^{x^2}-1} = \frac{\sqrt{\pi}}{4}\left(1+\frac{1}{2^{\frac{3}{2}}}+\frac{1}{3^{\frac{3}{2}}}+\frac{1}{4^{\frac{3}{2}}}+\ldots\right)$$

$$= \frac{\sqrt{\pi}}{4}\zeta(\tfrac{3}{2}) = \frac{\sqrt{\pi}}{4}2{\cdot}612,$$

so

$$1 = \frac{(2\pi mkT)^{\frac{3}{2}}}{h^3}\frac{V}{n}2{\cdot}612. \tag{8·1''}$$

The strange thing is that this is the largest value the integral can reach for $\zeta \leqslant 1$. But remember the equation was set up to determine ζ from n and the other data. It was the equation for the minimum in the steepest descent method. And a minimum there certainly was in every case. Yet we are faced with the fact that, if at a given temperature in a given volume there are more particles than the amount n determined from (8·1″), we cannot determine ζ.

There is nothing for it but to go back to the original form of the equation which was (see (7·14))

$$n = \sum_s \frac{1}{\frac{1}{\zeta}e^{\mu\alpha_s}-1} \quad \left(\mu = \frac{1}{kT}\right). \tag{7·14}$$

There it is immediately clear that there is no upper limit to the sum. Whether the first, the lowest, α_s is exactly zero (as we have taken it to be) or not, the first term of the sum can be made as large as we please without any term becoming negative, if we let ζ approach the value of $e^{\mu\alpha_0}$ (whether the latter be exactly 1, or a little larger than 1) from below.

In order not to confuse the issue, let us keep to $\alpha_0 = 0$, $e^{\mu\alpha_0} = 1$, though it is irrelevant and quantum-mechanically not quite correct. By letting ζ take a value very near to 1, viz.

$$\zeta = 1 - \frac{1}{\beta n} \quad \text{with (say)} \quad n^{-\frac{1}{3}} < \beta < 1, \tag{8·36}$$

any considerable fraction, say βn, of the particles can be accommodated in the lowest level (the simple modification that is needed, when the first 2 or 5 or 20 levels should be exactly equal—degenerate—can be left to the reader).

Now what about the next term of the sum? In the lowest levels the product of the momentum* p and the dimensions of the container $V^{\frac{1}{3}}$ is of the order $h/2\pi$:

$$pV^{\frac{1}{3}} \sim \frac{h}{2\pi},$$

hence the energy
$$\alpha = \frac{p^2}{2m} \sim \frac{h^2}{8\pi^2 m} V^{-\frac{2}{3}}; \tag{8·37}$$

and this is also the order of magnitude of the lowest level steps, i.e. of the differences between successive α's in the lowest region. Hence for the term following α_0 we shall have

$$\mu\alpha = \frac{\alpha}{kT} \sim \frac{h^2}{8\pi^2 mkT} V^{-\frac{2}{3}}.$$

This is still very small, but not of the order n^{-1}, only $n^{-\frac{2}{3}}$, as can be seen from (8·1″), which will, of course, hold as to order of magnitude. Then from (7·14) and (8·36) we may in this next term already safely put $\zeta = 1$ and obtain an occupation number of the order of

$$n^{\frac{2}{3}}$$

which is large, yet only an infinitesimal fraction of n (compare

* Please do not mix up the p in the following few lines with the pressure.

with the first occupation number βn, which may be a considerable fraction of n). Not larger are the following occupation numbers, for they decrease monotonically. Skipping the next 50 or 1000, we soon come to a region where the relative change from one level to the next is also small enough to justify completely the approximation by an integral (in fact, our 'count' $\frac{4\pi V}{R^3} p^2 dp$ does not, in any event, do justice to the lowest region which is quite irrelevant except, in this present case, for the very lowest level or levels). And, of course, in the integral we may safely use $\zeta = 1$.

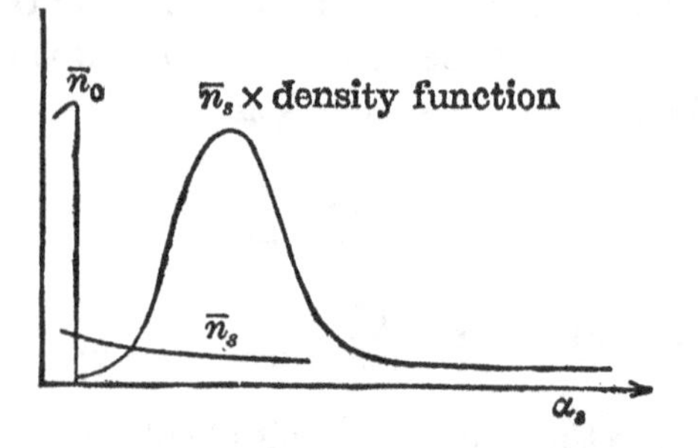

In a word, (8·1″) remains valid whenever the number of particles actually present is equal to, or greater than, the value of n which (8·1″) indicates. But only the number n will remain 'alive', as it were, spread over the energy line according to a law, akin to that of black-body radiation, while the surplus 'condenses', as it were, into the very lowest state (see figure).

On compression or dilatation, if we keep the temperature constant, the body will behave much like a saturated vapour in contact with its condensed state. The thermodynamic state (e.g. the pressure, the energy density) will not change until either everything is condensed or everything is evaporated (which means not that the lowest state is then not occupied at all, but that the 'hump' has disappeared).

The 'heat of evaporation' is, of course, precisely the mean

energy U of the particles in the 'alive' state; we take it from (8·4) with $\zeta = 1$:

$$\frac{2}{3}\frac{U}{nkT} = \frac{2}{3}\frac{\int_0^\infty \frac{x^4\,dx}{e^{x^2}-1}}{\int_0^\infty \frac{x^2\,dx}{e^{x^2}-1}}.$$

The integral in the denominator has already been indicated. The other is

$$\int_0^\infty \frac{x^4\,dx}{e^{x^2}-1} = \frac{3}{2}\frac{\sqrt{\pi}}{4}\left(1+\frac{1}{2^{\frac{5}{2}}}+\frac{1}{3^{\frac{5}{2}}}+\frac{1}{4^{\frac{5}{2}}}+\dots\right)$$

$$= \frac{3}{2}\frac{\sqrt{\pi}}{4}\zeta\left(\frac{5}{2}\right) = \frac{3}{2}\frac{\sqrt{\pi}}{4}1{\cdot}341.$$

Hence $$\frac{2}{3}\frac{U}{nkT} = \frac{1{\cdot}341}{2{\cdot}612} = 0{\cdot}5134.$$

We see that the energy in the saturated Bose-Einstein state has just a little more than half its classical value (the same holds for the pressure). If by isothermic compression we could clearly reach this state and go beyond it (which is certainly not the case, because the *ideal* laws are strongly distorted by the volume of the particles and their mutual forces) the particles would lose about half their energy gradually by a change of the distribution function, the other half abruptly by Bose-Einstein condensation.

In itself, the region of this strong degeneration is by no means outside the reach of experiment. For helium, for example, the required density is not yet that of the liquid if we take T as low as 1° K. (the computation is easy, if we use the result given in (8·9) and compare (8·8) with our present (8·1)).

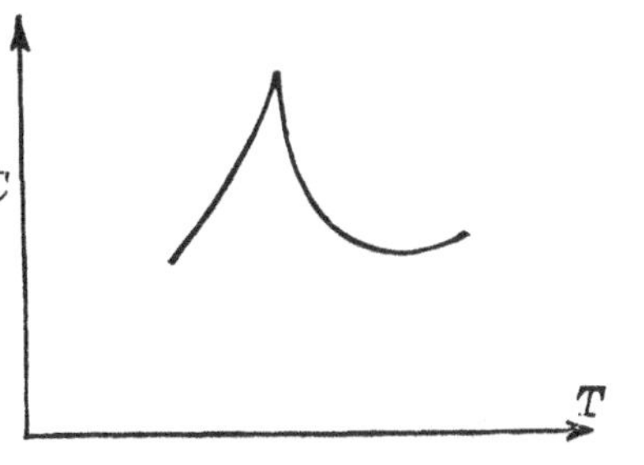

It has been pointed out by F. London that traces of this curious behaviour may be involved in the strange transition liquid helium exhibits around a certain low temperature of a few degrees K., the so-called Λ-point, where the specific heat, on cooling, shows a sudden rise and fall (of the shape of a Λ). This is, indeed, what we might expect if a sort of veiled allelomorphic transition took place, with the latent heat, instead of occurring suddenly at one definite temperature, spread over a small range. The fact that helium, at atmospheric pressure, remains liquid even at $T = 0$, indicates that this state, though liquid, must be 'peculiarly well ordered', if its entropy is to be zero.

Chapter IX

THE PROBLEM OF RADIATION*

At an early stage (p. 44) I drew attention to the fact that our considerations embrace the special case of electromagnetic radiation, which is characterized by

(i) vanishing rest-mass, which makes the energy α_s a linear function of the momentum p_s instead of a quadratic one as used hitherto;

(ii) $\zeta = 1$ or, in other words, an indefinite number of particles or quanta.

These two features are not mutually independent. For in the subsection on annihilation of matter we have shown that quanta with a non-vanishing rest-mass would practically disappear if their number were left indefinite.

There is, of course, some formal analogy with the case of Bose-Einstein condensation, where we also have $\zeta = 1$.

To obtain the customary theory, we need only envisage our general formula for the average occupation number in the Bose case, viz.

$$\bar{n}_s = \frac{1}{\frac{1}{\zeta}e^{\mu\alpha_s} - 1},$$

and put $\zeta = 1$ and $\alpha_s = h\nu_s$ (and of course $\mu = 1/kT$). Thus

$$\bar{n}_s = \frac{1}{e^{h\nu_s/kT} - 1} \tag{9·1}$$

is the average number of quanta $h\nu_s$ in the sth state (or on the sth 'hohlraum' oscillator). Considering that there are (see p. 49)

$$\frac{8\pi V\nu^2 d\nu}{c^3} \tag{9·2}$$

levels (or 'oscillators') with ν_s between ν and $\nu + d\nu$, we get for

* See note on page 88.

the energy density (i.e. with $V = 1$) of the monochromatic radiation between ν and $\nu + d\nu$

$$\frac{8\pi h}{c^3} \frac{\nu^3}{e^{h\nu/kT} - 1} d\nu, \tag{9·3}$$

which is the famous Planck formula.

But now one point calls for discussion. I have deliberately recalled in the preceding sentences the close analogy between a state labelled by 's' and capable of 'accommodating' one, two, three, ... quanta $h\nu_s$, and a quantum-mechanical oscillator. If this view (which has, in fact, historical priority) is adopted, the energy amounts $n_s h\nu_s$, or

$$0, \quad h\nu_s, \quad 2h\nu_s, \quad 3h\nu_s, \quad \dots, \tag{9·4}$$

acquire the meaning of the energy levels of this oscillator. Now in quantum mechanics the oscillator levels are not integral, but half-odd integral multiples of a unit (viz. of h times the classical frequency). This theoretical result has been confirmed by experiment in all cases that could be put to test. The theoretician can hardly refrain from asking: Would it make any difference if we replaced the level scheme (9·4) by

$$\tfrac{1}{2}h\nu_s, \quad \tfrac{3}{2}h\nu_s, \quad \tfrac{5}{2}h\nu_s, \quad \dots? \tag{9·5}$$

Now, this new assumption is not really covered by our formula (7·1)

$$Z = \sum_{(n_s)} e^{-\mu \sum_s n_s \alpha_s} \tag{7·1}$$

(to be summed over all permissible combinations $n_1, n_2, \dots, n_s, \dots$) because in all subsequent reasoning based upon it we had always taken the n_s to be integers. But it is easy to estimate the change it would involve. For in the general method of the 'sum-over-states' the absolute zero level of the energy is irrelevant. If a constant C is added to all the levels of the whole system, this constant drops out in all results—except, of course, that the average energy U is increased by this constant. Now in (9·5), as compared with (9·4), we have increased all the levels of the whole system by the constant

$$\sum_s \frac{h\nu_s}{2} = \frac{h}{2} \sum_s \nu_s. \tag{9·6}$$

It is true that this is an infinite constant. All we can really say is: If, at first, one introduces the innovation (9·5) up to $s = r$ (some large number) and, of course, in doing so, declares the zero-point energy

$$\frac{h}{2} \sum_{s=1}^{r} \nu_s$$

to be something that is always present and cannot therefore show up in any experiment on emission or absorption, then nothing is changed, however large one takes r; and one may perhaps consent to let $r \to \infty$. But there is little point in tormenting one's brain as to whether it is permissible or not to introduce this 'infinite zero-point energy'. The whole procedure is clearly a toy, a plaything introduced to satisfy the quantum physicist who fancies half-odd integral quantum numbers, rather than integral ones.

But an entirely new point of view has turned up in the recent work of Peng and Born, work that is designed to cope with much more serious difficulties which arise in the theory of radiation (and field quantization in general) when you go beyond contemplating the state of thermodynamical equilibrium and embark on the quantum-mechanical investigation of single individual processes of interaction. Whether Born and Peng's theory will really be successful in surmounting these difficulties cannot yet be said. Here I wish merely to indicate briefly their attitude towards the equilibrium problem.

Their theory leads them to attribute to any one of those 'hohlraum' oscillators (characterized by the label s) two fundamentally different situations (I say 'situations', because the term 'state' has already been used). It can either be not excited at all—when it has the energy zero—or excited, when it has one of the energies (9·5) and we might say that it accommodates 0, 1, 2, 3, ... quanta or 'atoms of radiation' (but we shall call its occupation numbers $\frac{1}{2}$, $\frac{3}{2}$, $\frac{5}{2}$, ... because that is simpler). As in the customary theory, the number of quanta remains *indefinite*. But the number of oscillators that are at all excited is to be a prescribed number, which we call N.

The theory must be built up anew, but it is very simple. We start from (7·1) in which we again put, as in (7·4),

$$e^{-\mu\alpha_s} = z_s, \quad \mu = \frac{1}{kT}; \tag{7·4}$$

thus

$$Z = \sum_{(n_s)} z_1^{n_1} z_2^{n_2} \ldots z^{n_s} \ldots. \tag{9·7}$$

The new assumptions amount to admitting for every n_s

$$n_s = 0, \quad \tfrac{1}{2}, \quad \tfrac{3}{2}, \quad \tfrac{5}{2}, \quad \ldots, \tag{9·8}$$

but with the accessory condition, that only and exactly N out of all the n_s's shall differ from zero.

In order to cope with this condition by complex integration, as we did in previous cases, we first attach to every power $z_r^{n_r}$ with $n_r \neq 0$ in (9·7) the factor ζ and then form the sum, disregarding the accessory condition. The result we call

$$\begin{aligned} f(\zeta) &= \prod_s (1 + \zeta z_s^{\frac{1}{2}} + \zeta z_s^{\frac{3}{2}} + \ldots) \\ &= \prod_s \left(1 + \zeta z_s^{\frac{1}{2}} \frac{1}{1 - z_s}\right) \\ &= \prod_s \left(1 + \frac{\zeta}{z_s^{-\frac{1}{2}} - z_s^{\frac{1}{2}}}\right) \\ &= \prod_s \left(1 + \frac{\zeta}{2\sinh\left(\frac{\mu\alpha_s}{2}\right)}\right). \end{aligned} \tag{9·9}$$

Our Z is obviously the coefficient of ζ^N in $f(\zeta)$. Hence by a process, of which we have twice given the details, we get

$$\log Z = -(N+1)\log\zeta + \log f(\zeta), \tag{9·10}$$

where ζ is the real positive root of

$$0 = -\frac{N+1}{\zeta} + \frac{f'(\zeta)}{f(\zeta)}. \tag{9·11}$$

(In (9·10) a certain correction term is omitted right away. The reader will easily justify this for himself, if he so desires.) From the last expression (9·9) and from (9·11), where we drop the odd

'1' in '$N+1$', we easily find

$$N = \sum_s \frac{1}{\frac{2}{\zeta}\sinh\frac{\mu\alpha_s}{2}+1}. \tag{9·12}$$

It is easy to guess what the summandus means: since N is the number of excited oscillators, the term is probably the mean excitation number (as opposed to the mean occupation number $\bar{n}_s$).

We shall compute both beginning with the latter, because it is to us the more familiar thing. Contemplate (9·7), the sum-over-states. Its single terms are, as we know, the relative probabilities of the various possible states of the whole (each state characterized by a set of numbers n_s). $\bar{n}_s$, for a particular s, is found by multiplying every term by its n_s, summing, and dividing by Z itself. That can be done in this way:*

$$\bar{n}_s = z_s\frac{\partial \log Z}{\partial z_s} = -\frac{1}{\mu}\frac{\partial \log Z}{\partial \alpha_s}. \tag{9·13}$$

Now use (9·10) for $\log Z$. According to (9·11) the implicit ('via ζ') differentiation contributes nothing and from (9·9)

$$\log f(\zeta) = \sum_s \log\left(1+\frac{\zeta}{2\sinh\left(\frac{\mu\alpha_s}{2}\right)}\right). \tag{9·14}$$

Hence

$$\bar{n}_s = -\frac{1}{\mu}\frac{\partial \log Z}{\partial \alpha_s} = \frac{1}{1+\frac{\zeta}{2\sinh\left(\frac{\mu\alpha_s}{2}\right)}}\,\frac{\zeta}{2\left[\sinh\left(\frac{\mu\alpha_s}{2}\right)\right]^2}\,\tfrac{1}{2}\cosh\frac{\mu\alpha_s}{2}$$

$$= \frac{1}{\frac{2}{\zeta}\sinh\left(\frac{\mu\alpha_s}{2}\right)+1}\,\tfrac{1}{2}\operatorname{cotanh}\frac{\mu\alpha_s}{2}$$

$$= \frac{1}{\frac{2}{\zeta}\sinh\left(\frac{\mu\alpha_s}{2}+1\right)}\left(\frac{1}{2}+\frac{1}{e^{\mu\alpha_s}-1}\right). \tag{9·15}$$

The expression is translucent—but we delay its discussion.

* The expression is known from (7·2), and we might have quoted it from there.

The mean excitation number of the sth oscillator (call it $\bar{e}_s$) is obtained from (9·7) by taking the sum of all those terms in which $n_s \neq 0$ and dividing it by Z itself. Since the terms in question vanish for $z_s = 0$, while the others are not affected, we have

$$\bar{e}_s = \frac{Z - Z_{(z_s=0)}}{Z} = 1 - \frac{Z_{(z_s=0)}}{Z}.$$

Hence $$\log(1-\bar{e}_s) = \log Z_{(z_s=0)} - \log Z.$$

Since $z_s = 0$ means $\alpha_s = \infty$, we get from (9·10) and (9·9)

$$\log(1-\bar{e}_s) = -\log\left(1 + \frac{\zeta}{2\sinh\left(\frac{\mu\alpha_s}{2}\right)}\right), \tag{9·16}$$

or

$$\bar{e}_s = 1 - \frac{1}{1 + \dfrac{\zeta}{2\sinh\left(\frac{\mu\alpha_s}{2}\right)}}$$

$$= \frac{1}{\frac{2}{\zeta}\sinh\frac{\mu\alpha_s}{2} + 1}. \tag{9·17}$$

This is, indeed, as we anticipated, the summandus in (9·12) and also the first factor in (9·15).

The discussion is now very simple. The excitation numbers form something very like a Fermi distribution, except that $2\sinh\frac{\mu\alpha_s}{2}$ stands for $e^{\mu\alpha_s}$, which makes little difference. If we want (9·15) to represent virtually the Planck formula, the Fermi distribution $\bar{e}_s$ must be strongly degenerate, i.e. ζ must be very large. According to (9·12) this is attained by taking N very large. Then $\bar{e}_s$ will be very nearly 1 up to a certain s ($\approx N$), where it drops to zero. And then (9·15), which can be written

$$\bar{n}_s = \bar{e}_s\left(\frac{1}{2} + \frac{1}{e^{\mu\alpha_s} - 1}\right), \tag{9·18}$$

does not appreciably differ from the Planck distribution, provided that N is large enough to embrace virtually the whole

Planck distribution. (What is stopped near $s \approx N$ is only the contribution $\frac{1}{2}h\nu_s$ to the zero-point energy, which therefore remains finite.)

Well, that is the whole story, for the time being, except for one pertinent remark.

What stops the zero-point energy from becoming infinite is not the N-condition, but the admittance of the level 0 for each oscillator. The N-condition is required to restrict the total number of excited levels below, not above ('nach unten, nicht nach oben'). Indeed, nothing is changed if it is replaced by the inequality: number of excited levels to be $\geqslant N$ (but $\leqslant N$ would not do).

To see this, remember that the coefficient of ζ^N in $f(\zeta)$ was isolated by forming the residue of

$$\zeta^{-N-1} f(\zeta), \tag{9·19}$$

that this leads to (9·10), and that ζ must result very large. Now if we chose the inequality condition ($\geqslant N$) instead, we should have to collect the coefficients of all $\zeta^{N'}$ for which $N' \geqslant N$. But to avoid a second infinite process, let us rather collect those with $N' < N$ and subtract them from $f(1)$ (which is the sum of all coefficients). So we form

$$(\zeta^{-1} + \zeta^{-2} + \ldots + \zeta^{-N}) f(\zeta) = \zeta^{-1} \frac{1-\zeta^{-N}}{1-\zeta^{-1}} f(\zeta) = \frac{1-\zeta^{-N}}{\zeta - 1} f(\zeta)$$

and get for Z in the present case

$$Z = f(1) - \frac{1}{2\pi i} \oint \frac{1-\zeta^{-N}}{\zeta - 1} f(\zeta)\, d\zeta.$$

Note that the integrand has no singularity at $\zeta = 1$ (in fact none but the one at the origin); hence we may choose for integration a circle with $|\zeta| > 1$, and as large as we please. After that, we split the integral into the sum of two integrals, according to the numerator. The first, according to Cauchy's theorem, cancels $f(1)$ and we have

$$Z = \frac{1}{2\pi i} \oint \frac{\zeta^{-N}}{\zeta - 1} f(\zeta)\, d\zeta.$$

The fact that this now *has* a singularity at $\zeta = 1$ is irrelevant, for the equation is established; we have only to evaluate it. We do so by the steepest descent method. But then it is clear that with N a very large number

(i) the saddle-point will be found at $\zeta > 1$, provided the situation is such that it would be found at a ζ appreciably > 1 if the factor were ζ^{-N-1} in lieu of $\zeta^{-N}/\zeta - 1$;

(ii) that being so, the result will practically be the same as before (the change can hardly be greater, than if N changed by one unit).

If the N-condition were dropped altogether, one would, of course, get $\zeta = 1$ as on previous occasions. That is inadmissible, because it is easily seen that the result would deviate widely from Planck's formula.

Note. When this little book appeared first (1946) the idea of Peng and Born was quite recent. Since then I have sometimes thought of removing the whole of Chapter IX, in order that the reader should not take it too seriously. But now I definitely think it should stand (without suggesting that the idea be necessarily accepted) because it does yield an interesting insight into the probability structure of the equilibrium of radiation, an insight not easily obtained otherwise.

Appendix

THE CANONICAL DISTRIBUTION OF QUANTUM-MECHANICAL AMPLITUDES

The determination of the statistical entropy of a thermodynamical system always rests on counting the number of permutations that comply with certain restrictions, or, speaking in terms of physics, the number of different microstates that do not differ for the macroscopic observer because they all agree with regard to the macroscopic properties which alone can be observed by him, and among which usually the total energy plays a prominent role. For performing this count we have, in this book, always adopted the customary method, which rests on the assumption that every system, whether one of the big ones on which we actually experiment, or one of the innumerable microsystems of which we may consider the big one composed, always finds itself in a state of sharply defined energy; thus on one of its 'quantum levels of energy', if the system is governed by quantum mechanics; and this is, of course, the only case that interests us eventually.

In the beginning of Chapter II it was mentioned that this assumption is irreconcilable with the very foundations of quantum mechanics. We yet decided to adopt it, because it is the customary one, is very convenient, and gives essentially the same results as the consistent view, which must admit that the energy is as a rule never sharp but exhibits a certain spread, which is relatively (i.e. compared with the energy itself) small in big systems, but very often large in small systems. In fact we shall see that for a system in a heat bath this spread in the single system is very much the same as the statistical fluctuation in time according to the simplifying inconsistent view.

For I wish to supply here a general proof, that the consistent attitude leads to the same thermodynamical results as the convenient short-cut of putting every single microsystem on one of its sharp energy levels. This is important. For whatever one may think of the tenet of orthodox quantum theorists that by measuring the energy of a system we actually and *physically* 'put' it on to one of these levels, this tenet hardly justifies our doing so *in our minds* with the innumerable microsystems whose energy contents we do not measure, nay cannot even think of measuring. Taken literally, this would mean that a physical process, even when no observing 'subject' interferes with it, consists of continual sequences of fits and jerks, the successive transfers of energy parcels between microsystems. This view, when given serious thought, cannot pass for anything but a sometimes convenient metaphor.

We shall associate a wave function with the state of the system. We thereby ascribe it a 'pure state' as opposed to a 'mixture', the mathematical equivalent of which is a density matrix. A purist might challenge the use of a wave function not determined by measurement. But he would have to give up using wave functions altogether, since none has ever been determined by measurement. I see no objection to using it here, and it is sufficiently general to bring out our main point, namely, that the permutation numbers and the statistical entropy deduced from them follow directly from the scheme of eigenvalues of the energy; the customary but unjustifiable procedure of studying the possible distributions of an assembly over the allowed levels (or the like) is not needed.

We have to begin from scratch. The identification of statistical concepts in the model with thermodynamical concepts in the physical object always involves some basic assumptions. The quantum-energy-levels of big systems (on the laboratory scale) are highly degenerate. Let the multiplicity of the eigenvalue E_r of the energy be m_r. We assume that, in a big system, for the energy E_r the entropy S is

$$S = k \log m_r, \tag{A 1}$$

where k is Boltzmann's constant. This is, of course, in complete agreement with what we are used to, since m_r clearly is the 'number of different ways in which the energy E_r can be allocated in the system'. But for the present purpose we must call it a basic assumption (No. 1). The assumption No. 2 is more hazardous, it takes charge in the present context of the old crux of 'molecular disorder'. We assume that in a big system in certain circumstances (of small ill-defined perpetual disturbances; see below) the amplitude-squares or 'excitation strengths' are on the average equal for the m_r eigenfunctions belonging to the same eigenvalue E_r. This assumption is only required to hold in the very broad sense that in the long run there is no preference for any among those (as a rule innumerable) degenerate functions; moreover, it is quantum mechanically invariant, and in a special case it has yielded correct results for the intensities of the spectral components of the Stark-effect, where the multiplicity is small and the circumstances of the excitation would seem less balanced than in the thermodynamical application.

We now envisage, besides the system with energy-eigenvalues E_r of multiplicities m_r, a second system with energy-eigenvalues F_r of multiplicities p_r. We couple the two, but at first only in thought, just *regarding* them as one system, which we call 'the combined system'. Its general eigenvalue

$$E_r + F_s \qquad \text{(A 2)}$$

has at least the multiplicity $m_r p_s$ since any one of the m_r eigenfunctions of E_r (call it ψ for the moment) multiplied by any one of the p_s eigenfunctions of F_s (say ϕ) is an eigenfunction of the eigenvalue $E_r + F_s$ of the combined system. Moreover, in any state of the latter the excitation strength (absolute square of the vibrational amplitude) of any such product $\phi\psi$ contributes additively to the excitation strength of E_r in the first system, and, of course, to that of F_s in the second system. We now proceed as follows:

(i) we let the two systems be loosely coupled, but continue to describe them as before, i.e. as if they were not coupled; this is a permissible approximation, continually used in quantum mechanics;

(ii) we let the combined system be in its energy state

$$E = E_r + F_s; \tag{A 3}$$

(iii) we take the second system to be so vast (heat bath!) that for any E_r there is an F_s satisfying (A 3) with the given E. The multiplicity of E is therefore

$$\sum_r m_r p_s, \tag{A 4}$$

the sum to be taken over all r, with s the index for which (A 3) is satisfied. (Strictly speaking one ought to impose an upper limit on E_r; but in the limit of a very large heat bath this becomes irrelevant.)

From assumption No. 2, applied to the combined system, the total excitation strengths of those of its proper modes in which $E_1, E_2, E_3, \ldots, E_r, \ldots$ are involved, respectively, bear to each other the ratios of the products

$$m_1 p_{s'},\ m_2 p_{s''},\ m_3 p_{s'''},\ \ldots,\ m_r p_s \ldots, \tag{A 5}$$

where the notation, so I hope, is unambiguous. Hence these products also indicate the relative excitation strengths of the various E_r in the first system. Moreover, according to our assumption No. 1

$$k \log p_s = S(F_s) = S(E - E_r), \tag{A 6}$$

where S is the entropy of the heat bath as a function of its energy. Since E_r may be considered small compared with E

$$k \log p_s = S(E) - \frac{\partial S}{\partial E} E_r = S(E) - \frac{E_r}{T}, \tag{A 7}$$

where T is the temperature. Hence

$$m_r p_s = m_r e^{-E_r/kT} . e^{S(E)/k}. \tag{A 8}$$

The last factor, being independent of r, may be dropped, since we were only concerned with the ratios. The absolute values are a matter of normalization, the sum of *all* absolute amplitude squares for the first system alone being customarily normalized to unity.

We thus get in a heat bath of temperature T for our system (E_r), which may be a big or a small one, exactly the same canonical distribution between the amplitude squares, as is in the customary treatment said to indicate the probability of the system being on this or that energy level. Indeed, he who adheres to the orthodox tenets of quantum mechanics may find little difference between the two statements, since to him the amplitude square indicates precisely the probability of *finding* the system at the level in question if he measures its energy; while if he does not measure it, then, so he says, it does not concern him. With the latter I cannot agree. In the theoretical analysis of a concrete experiment we usually have to consider a great many physical data which are not subject to a measuring process in that particular experiment.

To illustrate the greater economy, as regards basic assumptions, of the method of thought advocated here, let us take for our system (E_r) the electromagnetic radiation in a cavity, surrounded by thick walls of uniform temperature. It will be found that Planck's formula for black body radiation can be deduced merely by ascribing to each 'radiation oscillator' (i.e. classical proper mode of the cavity) the well-known equidistant levels of a quantum oscillator and considering the levels of the whole 'body of radiation' additively composed of the oscillator levels in all combinations. I say, merely this level structure is needed and is relevant. It is not necessary to make the much more incisive assumption which is usually made, viz. that each individual radiation oscillator always carries an integral number of quanta $h\nu$, or as is sometimes said, that there is always an integral number of photons of that particular brand in the state indicated by that individual radiation oscillator. The concept of photons or 'parcels of radiative energy'—beyond the fundamental notion of the structure of the spectrum of eigenvalues—becomes gratuitous, anyhow for the purpose of statistical thermodynamics.

There is one inconsistency in our derivation of the 'canonical distribution' (A 8), namely that the combined system (heat bath + the system under consideration) was assumed to contain a definite amount of energy E. Our deduction follows the old pattern of starting with a 'microcanonical distribution' of a big system, and obtaining a canonical distribution for a small part of it that is loosely coupled to the rest. I do not think that this traditional inconsistency is very serious. It is quite obvious that a moderately sharp (e.g. a canonical) distribution of the combined system leads to the same result.

From the canonical distribution of amplitudes, deduced in (A 8), the complete thermodynamics of the system can be derived. To include 'external work', one has to let the levels E_r depend on observable parameters, as was explained on pp. 11 ff. of the main text. We need not repeat this here, but I wish to show that our basic assumption (A 1) for big systems may be *a posteriori*, if not justified, at least shown to be consistent with what follows from (A 8). The mean value of the energy is

$$\bar{E}_r = \frac{\Sigma E_r m_r e^{-E_r/kT}}{\Sigma m_r e^{-E_r/kT}} = kT^2 \frac{d \log Z}{dT}, \tag{A 9}$$

where

$$Z = \Sigma m_r e^{-E_r/kT} \tag{A 10}$$

is the 'sum-over-states'. Thus

$$\bar{E}_r dT = kT^2 d \log Z$$

or

$$-d\left(\frac{\bar{E}_r}{T}\right) + \frac{d\bar{E}_r}{T} = kd \log Z,$$

and finally

$$d\bar{E}_r = Td\left(\frac{\bar{E}_r}{T} + k \log Z\right). \tag{A 11}$$

It is therefore consistent to put the entropy S *of our system*

$$S = \frac{\bar{E}_r}{T} + k \log Z = k \frac{d}{dT}(T \log Z). \tag{A 12}$$

If the system is very big, the sum-over-states may be replaced by its maximum term, and $\bar{E}_r$ by the 'maximum' E_r. Then we get from (A 10) and the first equation (A 12)—the index r now referring to the maximum—

$$S = k \log m_r, \tag{A 13}$$

which agrees with (A 1). Alternatively the maximum term in (A 10) for a given T is determined by differentiating the general term 'with respect to r' and equating the result to zero:

$$d\log m_r - \frac{1}{kT}\,dE_r = 0,$$

or

$$dE_r = Td(k\log m_r). \tag{A 14}$$

This agrees thermodynamically with (A 13), and the agreement is meaningful. For the relation (A 14) between the two increments holds for every T near the maximum term. Hence it must hold for the actual increments, when the maximum is shifted on account of a change of T.

A CATALOG OF SELECTED
DOVER BOOKS
IN SCIENCE AND MATHEMATICS

Astronomy

CHARIOTS FOR APOLLO: The NASA History of Manned Lunar Spacecraft to 1969, Courtney G. Brooks, James M. Grimwood, and Loyd S. Swenson, Jr. This illustrated history by a trio of experts is the definitive reference on the Apollo spacecraft and lunar modules. It traces the vehicles' design, development, and operation in space. More than 100 photographs and illustrations. 576pp. 6 3/4 x 9 1/4. 0-486-46756-2

EXPLORING THE MOON THROUGH BINOCULARS AND SMALL TELESCOPES, Ernest H. Cherrington, Jr. Informative, profusely illustrated guide to locating and identifying craters, rills, seas, mountains, other lunar features. Newly revised and updated with special section of new photos. Over 100 photos and diagrams. 240pp. 8 1/4 x 11. 0-486-24491-1

WHERE NO MAN HAS GONE BEFORE: A History of NASA's Apollo Lunar Expeditions, William David Compton. Introduction by Paul Dickson. This official NASA history traces behind-the-scenes conflicts and cooperation between scientists and engineers. The first half concerns preparations for the Moon landings, and the second half documents the flights that followed Apollo 11. 1989 edition. 432pp. 7 x 10. 0-486-47888-2

APOLLO EXPEDITIONS TO THE MOON: The NASA History, Edited by Edgar M. Cortright. Official NASA publication marks the 40th anniversary of the first lunar landing and features essays by project participants recalling engineering and administrative challenges. Accessible, jargon-free accounts, highlighted by numerous illustrations. 336pp. 8 3/8 x 10 7/8. 0-486-47175-6

ON MARS: Exploration of the Red Planet, 1958-1978--The NASA History, Edward Clinton Ezell and Linda Neuman Ezell. NASA's official history chronicles the start of our explorations of our planetary neighbor. It recounts cooperation among government, industry, and academia, and it features dozens of photos from Viking cameras. 560pp. 6 3/4 x 9 1/4. 0-486-46757-0

ARISTARCHUS OF SAMOS: The Ancient Copernicus, Sir Thomas Heath. Heath's history of astronomy ranges from Homer and Hesiod to Aristarchus and includes quotes from numerous thinkers, compilers, and scholasticists from Thales and Anaximander through Pythagoras, Plato, Aristotle, and Heraclides. 34 figures. 448pp. 5 3/8 x 8 1/2. 0-486-43886-4

AN INTRODUCTION TO CELESTIAL MECHANICS, Forest Ray Moulton. Classic text still unsurpassed in presentation of fundamental principles. Covers rectilinear motion, central forces, problems of two and three bodies, much more. Includes over 200 problems, some with answers. 437pp. 5 3/8 x 8 1/2. 0-486-64687-4

BEYOND THE ATMOSPHERE: Early Years of Space Science, Homer E. Newell. This exciting survey is the work of a top NASA administrator who chronicles technological advances, the relationship of space science to general science, and the space program's social, political, and economic contexts. 528pp. 6 3/4 x 9 1/4. 0-486-47464-X

STAR LORE: Myths, Legends, and Facts, William Tyler Olcott. Captivating retellings of the origins and histories of ancient star groups include Pegasus, Ursa Major, Pleiades, signs of the zodiac, and other constellations. "Classic." – *Sky & Telescope.* 58 illustrations. 544pp. 5 3/8 x 8 1/2. 0-486-43581-4

A COMPLETE MANUAL OF AMATEUR ASTRONOMY: Tools and Techniques for Astronomical Observations, P. Clay Sherrod with Thomas L. Koed. Concise, highly readable book discusses the selection, set-up, and maintenance of a telescope; amateur studies of the sun; lunar topography and occultations; and more. 124 figures. 26 halftones. 37 tables. 335pp. 6 1/2 x 9 1/4. 0-486-42820-6

Chemistry

MOLECULAR COLLISION THEORY, M. S. Child. This high-level monograph offers an analytical treatment of classical scattering by a central force, quantum scattering by a central force, elastic scattering phase shifts, and semi-classical elastic scattering. 1974 edition. 310pp. 5 3/8 x 8 1/2. 0-486-69437-2

HANDBOOK OF COMPUTATIONAL QUANTUM CHEMISTRY, David B. Cook. This comprehensive text provides upper-level undergraduates and graduate students with an accessible introduction to the implementation of quantum ideas in molecular modeling, exploring practical applications alongside theoretical explanations. 1998 edition. 832pp. 5 3/8 x 8 1/2. 0-486-44307-8

RADIOACTIVE SUBSTANCES, Marie Curie. The celebrated scientist's thesis, which directly preceded her 1903 Nobel Prize, discusses establishing atomic character of radioactivity; extraction from pitchblende of polonium and radium; isolation of pure radium chloride; more. 96pp. 5 3/8 x 8 1/2. 0-486-42550-9

CHEMICAL MAGIC, Leonard A. Ford. Classic guide provides intriguing entertainment while elucidating sound scientific principles, with more than 100 unusual stunts: cold fire, dust explosions, a nylon rope trick, a disappearing beaker, much more. 128pp. 5 3/8 x 8 1/2. 0-486-67628-5

ALCHEMY, E. J. Holmyard. Classic study by noted authority covers 2,000 years of alchemical history: religious, mystical overtones; apparatus; signs, symbols, and secret terms; advent of scientific method, much more. Illustrated. 320pp. 5 3/8 x 8 1/2. 0-486-26298-7

CHEMICAL KINETICS AND REACTION DYNAMICS, Paul L. Houston. This text teaches the principles underlying modern chemical kinetics in a clear, direct fashion, using several examples to enhance basic understanding. Solutions to selected problems. 2001 edition. 352pp. 8 3/8 x 11. 0-486-45334-0

PROBLEMS AND SOLUTIONS IN QUANTUM CHEMISTRY AND PHYSICS, Charles S. Johnson and Lee G. Pedersen. Unusually varied problems, with detailed solutions, cover of quantum mechanics, wave mechanics, angular momentum, molecular spectroscopy, scattering theory, more. 280 problems, plus 139 supplementary exercises. 430pp. 6 1/2 x 9 1/4. 0-486-65236-X

ELEMENTS OF CHEMISTRY, Antoine Lavoisier. Monumental classic by the founder of modern chemistry features first explicit statement of law of conservation of matter in chemical change, and more. Facsimile reprint of original (1790) Kerr translation. 539pp. 5 3/8 x 8 1/2. 0-486-64624-6

MAGNETISM AND TRANSITION METAL COMPLEXES, F. E. Mabbs and D. J. Machin. A detailed view of the calculation methods involved in the magnetic properties of transition metal complexes, this volume offers sufficient background for original work in the field. 1973 edition. 240pp. 5 3/8 x 8 1/2. 0-486-46284-6

GENERAL CHEMISTRY, Linus Pauling. Revised third edition of classic first-year text by Nobel laureate. Atomic and molecular structure, quantum mechanics, statistical mechanics, thermodynamics correlated with descriptive chemistry. Problems. 992pp. 5 3/8 x 8 1/2. 0-486-65622-5

ELECTROLYTE SOLUTIONS: Second Revised Edition, R. A. Robinson and R. H. Stokes. Classic text deals primarily with measurement, interpretation of conductance, chemical potential, and diffusion in electrolyte solutions. Detailed theoretical interpretations, plus extensive tables of thermodynamic and transport properties. 1970 edition. 590pp. 5 3/8 x 8 1/2. 0-486-42225-9

Engineering

FUNDAMENTALS OF ASTRODYNAMICS, Roger R. Bate, Donald D. Mueller, and Jerry E. White. Teaching text developed by U.S. Air Force Academy develops the basic two-body and n-body equations of motion; orbit determination; classical orbital elements, coordinate transformations; differential correction; more. 1971 edition. 455pp. 5 3/8 x 8 1/2. 0-486-60061-0

INTRODUCTION TO CONTINUUM MECHANICS FOR ENGINEERS: Revised Edition, Ray M. Bowen. This self-contained text introduces classical continuum models within a modern framework. Its numerous exercises illustrate the governing principles, linearizations, and other approximations that constitute classical continuum models. 2007 edition. 320pp. 6 1/8 x 9 1/4. 0-486-47460-7

ENGINEERING MECHANICS FOR STRUCTURES, Louis L. Bucciarelli. This text explores the mechanics of solids and statics as well as the strength of materials and elasticity theory. Its many design exercises encourage creative initiative and systems thinking. 2009 edition. 320pp. 6 1/8 x 9 1/4. 0-486-46855-0

FEEDBACK CONTROL THEORY, John C. Doyle, Bruce A. Francis and Allen R. Tannenbaum. This excellent introduction to feedback control system design offers a theoretical approach that captures the essential issues and can be applied to a wide range of practical problems. 1992 edition. 224pp. 6 1/2 x 9 1/4. 0-486-46933-6

THE FORCES OF MATTER, Michael Faraday. These lectures by a famous inventor offer an easy-to-understand introduction to the interactions of the universe's physical forces. Six essays explore gravitation, cohesion, chemical affinity, heat, magnetism, and electricity. 1993 edition. 96pp. 5 3/8 x 8 1/2. 0-486-47482-8

DYNAMICS, Lawrence E. Goodman and William H. Warner. Beginning engineering text introduces calculus of vectors, particle motion, dynamics of particle systems and plane rigid bodies, technical applications in plane motions, and more. Exercises and answers in every chapter. 619pp. 5 3/8 x 8 1/2. 0-486-42006-X

ADAPTIVE FILTERING PREDICTION AND CONTROL, Graham C. Goodwin and Kwai Sang Sin. This unified survey focuses on linear discrete-time systems and explores natural extensions to nonlinear systems. It emphasizes discrete-time systems, summarizing theoretical and practical aspects of a large class of adaptive algorithms. 1984 edition. 560pp. 6 1/2 x 9 1/4. 0-486-46932-8

INDUCTANCE CALCULATIONS, Frederick W. Grover. This authoritative reference enables the design of virtually every type of inductor. It features a single simple formula for each type of inductor, together with tables containing essential numerical factors. 1946 edition. 304pp. 5 3/8 x 8 1/2. 0-486-47440-2

THERMODYNAMICS: Foundations and Applications, Elias P. Gyftopoulos and Gian Paolo Beretta. Designed by two MIT professors, this authoritative text discusses basic concepts and applications in detail, emphasizing generality, definitions, and logical consistency. More than 300 solved problems cover realistic energy systems and processes. 800pp. 6 1/8 x 9 1/4. 0-486-43932-1

THE FINITE ELEMENT METHOD: Linear Static and Dynamic Finite Element Analysis, Thomas J. R. Hughes. Text for students without in-depth mathematical training, this text includes a comprehensive presentation and analysis of algorithms of time-dependent phenomena plus beam, plate, and shell theories. Solution guide available upon request. 672pp. 6 1/2 x 9 1/4. 0-486-41181-8

HELICOPTER THEORY, Wayne Johnson. Monumental engineering text covers vertical flight, forward flight, performance, mathematics of rotating systems, rotary wing dynamics and aerodynamics, aeroelasticity, stability and control, stall, noise, and more. 189 illustrations. 1980 edition. 1089pp. 5 5/8 x 8 1/4. 0-486-68230-7

MATHEMATICAL HANDBOOK FOR SCIENTISTS AND ENGINEERS: Definitions, Theorems, and Formulas for Reference and Review, Granino A. Korn and Theresa M. Korn. Convenient access to information from every area of mathematics: Fourier transforms, Z transforms, linear and nonlinear programming, calculus of variations, random-process theory, special functions, combinatorial analysis, game theory, much more. 1152pp. 5 3/8 x 8 1/2. 0-486-41147-8

A HEAT TRANSFER TEXTBOOK: Fourth Edition, John H. Lienhard V and John H. Lienhard IV. This introduction to heat and mass transfer for engineering students features worked examples and end-of-chapter exercises. Worked examples and end-of-chapter exercises appear throughout the book, along with well-drawn, illuminating figures. 768pp. 7 x 9 1/4. 0-486-47931-5

BASIC ELECTRICITY, U.S. Bureau of Naval Personnel. Originally a training course; best nontechnical coverage. Topics include batteries, circuits, conductors, AC and DC, inductance and capacitance, generators, motors, transformers, amplifiers, etc. Many questions with answers. 349 illustrations. 1969 edition. 448pp. 6 1/2 x 9 1/4. 0-486-20973-3

BASIC ELECTRONICS, U.S. Bureau of Naval Personnel. Clear, well-illustrated introduction to electronic equipment covers numerous essential topics: electron tubes, semiconductors, electronic power supplies, tuned circuits, amplifiers, receivers, ranging and navigation systems, computers, antennas, more. 560 illustrations. 567pp. 6 1/2 x 9 1/4. 0-486-21076-6

BASIC WING AND AIRFOIL THEORY, Alan Pope. This self-contained treatment by a pioneer in the study of wind effects covers flow functions, airfoil construction and pressure distribution, finite and monoplane wings, and many other subjects. 1951 edition. 320pp. 5 3/8 x 8 1/2. 0-486-47188-8

SYNTHETIC FUELS, Ronald F. Probstein and R. Edwin Hicks. This unified presentation examines the methods and processes for converting coal, oil, shale, tar sands, and various forms of biomass into liquid, gaseous, and clean solid fuels. 1982 edition. 512pp. 6 1/8 x 9 1/4. 0-486-44977-7

THEORY OF ELASTIC STABILITY, Stephen P. Timoshenko and James M. Gere. Written by world-renowned authorities on mechanics, this classic ranges from theoretical explanations of 2- and 3-D stress and strain to practical applications such as torsion, bending, and thermal stress. 1961 edition. 560pp. 5 3/8 x 8 1/2. 0-486-47207-8

PRINCIPLES OF DIGITAL COMMUNICATION AND CODING, Andrew J. Viterbi and Jim K. Omura. This classic by two digital communications experts is geared toward students of communications theory and to designers of channels, links, terminals, modems, or networks used to transmit and receive digital messages. 1979 edition. 576pp. 6 1/8 x 9 1/4. 0-486-46901-8

LINEAR SYSTEM THEORY: The State Space Approach, Lotfi A. Zadeh and Charles A. Desoer. Written by two pioneers in the field, this exploration of the state space approach focuses on problems of stability and control, plus connections between this approach and classical techniques. 1963 edition. 656pp. 6 1/8 x 9 1/4. 0-486-46663-9